APPLIED MATHEMATICS
Essential Theory and Exercises

P. J. Holt BA, BSc, PhD
Head of Mathematics, Wimbledon College

HODDER AND STOUGHTON
LONDON SYDNEY AUCKLAND TORONTO

British Library Cataloguing in Publication Data

Holt, P. J.
   Applied mathematics: essential theory and
   exercises.—2nd ed.
   1. Mechanics
   I. Title   II. Holt, P. J. Mechanics
   531   QC127

   ISBN 0 340 37158 7

First printed as *Mechanics: Essential Theory and Exercises* 1980
Second edition 1985

Typeset by Macmillan India Ltd., Bangalore
Printed in Great Britain for Hodder and Stoughton Educational,
a division of Hodder and Stoughton Ltd,
Mill Road, Dunton Green, Sevenoaks, Kent, by
Richard Clay (The Chaucer Press), Ltd,
Bungay, Suffolk

Contents

Preface

This book is a new revised edition of *Mechanics: Essential Theory and Exercises* which was published in 1980. In writing it I set out to provide a complete but inexpensive textbook with the emphasis on short but clear explanations, worked examples and carefully graded exercises containing plenty of *numerical* problems for the benefit of the many students who initially find mechanics difficult. The book has now been used in my own department for four years, and while it has proved popular and successful, I have inevitably discovered a few deficiencies and areas needing attention.

Perhaps the major improvement in the new edition is the addition of a chapter on probability, which means that the book now covers the *applied mathematics* (rather than simply the mechanics) which is required for most single subject mathematics A level examinations. Another important addition is a set of revision exercises, each covering the material in one particular chapter. I believe these to be essential since the topics in applied mathematics are very seldom mastered at the first attempt. Finally a number of the original exercises have been re-written, mainly because I have found that even more simple numerical examples were required.

I would like to take this opportunity of thanking my friend and ex-colleague Mr Bill Doris, who has used the book in his own school and made many helpful comments and suggestions.

Preliminary Notes

SI units

In this book we shall work almost entirely in the system of units called the *Système International* (SI), which has been introduced to bring about international conformity in the use of scientific units. A brief explanation of the system may be helpful.

There are three fundamental quantities in mechanics, in terms of which all other quantities are defined. These are *mass*, *length* and *time*, and their SI units – known as base units – are the *kilogram* (kg), the *metre* (m) and the *second* (s). The SI units of other quantities, which are defined in terms of the fundamental ones, are called derived units, and they are formed from the base units by multiplication and division in accordance with their definitions.

Example There is a quantity in mechanics called linear *momentum*, which is defined to be the mass of a body multiplied by its velocity. Analysing this into the three fundamental quantities we have

$$\text{momentum} = \text{a mass} \times \text{a distance} \div \text{a time}.$$

To obtain the SI unit of momentum we replace each fundamental quantity in the definition by its base unit, and link the units by using a small gap to indicate multiplication and a negative index to indicate division:

$$\mathbf{kg\ m\ s^{-1}}$$

A solidus (stroke) can also be used to indicate division, but its use is normally limited to pre-A-Level work.

Decimal multiples of SI units

Although there is only one SI unit of each physical quantity, which is derived from the three base units as explained, decimal multiples of the SI units may be formed by attaching prefixes to them. The following are some of the more commonly used prefixes.

Multiple	Prefix	Symbol	Multiple	Prefix	Symbol
10^{-2}	centi	c	10^{3}	kilo	k
10^{-3}	milli	m	10^{6}	mega	M

Thus for example $1000 \text{ m} = 1$ kilometre or 1 km.

Some of the more complicated SI units have special names. For example the unit of force, $kg\,m\,s^{-2}$, is called the *newton*. The accompanying table gives the SI units used in this book, together with their names and symbols. The principal non-SI unit used in the book is the *hour*, denoted by h.

Quantity	SI unit and decimal multiples	Quantity	SI unit and decimal multiples
Mass	kilogram, kg gram, g megagram, Mg – also called the *tonne*	Density Angle	kilogram per metre cubed, $kg\,m^{-3}$ radian, rad degree, °
Length	metre, m centimetre, cm kilometre, km	Angular speed Force	radian per second, $rad\,s^{-1}$ newton, N $(= kg\,m\,s^{-2})$
Time	second, s		kilonewton, kN
Speed and velocity	metre per second, $m\,s^{-1}$	Moment	newton metre, N m
Acceleration	metre per second squared, $m\,s^{-2}$	Momentum, impulse	N s or $kg\,m\,s^{-1}$
Area	square metre, m^2 square centimetre, cm^2	Work, energy	joule, J $(= Nm)$ kilojoule, kJ
Volume	cubic metre, m^3	Power	watt, W $(=$ joule per second) kilowatt, kW

Symbols and conventions used in this book

Throughout the book the acceleration due to gravity, g, is taken to be $10\,m\,s^{-2}$. When it arises, the density of water is taken to be $1000\,kg\,m^{-3}$.

A box around a prominent statement means (a) that the statement should be learned, (b) that it is quotable without proof unless a proof is specifically required. Double arrow-heads, $\longrightarrow\!\!\!\rightarrow$, have two uses. In problems involving moving bodies they indicate *accelerations* and thus enable accelerations to be distinguished from forces and velocities. In diagrams representing systems of forces or other vector quantities they indicate *resultants* when it is necessary to show such resultants in the same diagrams as the systems they replace. No diagrams occur in which both accelerations and resultants need to be shown.

In worked examples units are usually omitted until the answer is obtained, unless this could cause confusion. Answers to both worked examples and exercises are normally given to four significant figures unless a form such as $\sqrt{3}$, arc tan 2, etc. is appropriate. Very elementary algebraic steps are omitted in the worked examples, as are arithmetical calculations which could appropriately be performed by calculator. Angles are expressed in decimals of a degree rather than minutes.

1

Motion with Constant Acceleration

Quantities associated with motion

Suppose a small body is moving along a straight line. We can specify its position at any given time by giving its distance from some fixed point on the line and the direction in which this distance is measured – e.g. 'to the right' or 'to the left'. Such a 'directed distance' is called a *displacement* and denoted by *s*. It is a simple example of a *vector* quantity, that is one having both magnitude and direction. When we are dealing with motion in a straight line only two directions are possible and it is convenient to take one of them as positive and the other as negative. Then we can distinguish between the two directions by means of plus and minus signs. For a body moving along an *x*-axis, therefore, which has accepted positive and negative directions, the displacement from the origin is − 2 when the body is at the point (− 2,0), and + 5 (or simply 5) when it is at (5,0).

Another important property of a moving body is its *velocity*, *v*. In everyday language this term is virtually synonymous with *speed*; but in the strict usage of mechanics velocity is related to speed just as displacement is related to distance: a velocity is a speed *in a definite direction*. For a body moving in a straight line we can again distinguish between directions by means of plus and minus signs; so that, for example, a body moving along an *x*-axis at 8 m s^{-1} to the left would be said to have a velocity of − 8 m s^{-1}.

Somewhat less familiar is the quantity *acceleration* (*a*). This is the rate at which velocity is being gained, or, more simply, the velocity gained per second or per hour (for example). The SI unit is the *metre per second per second*, written as m s^{-2}. Like displacement and velocity, acceleration is a vector quantity, its direction being that in which velocity is being gained.

Example Suppose a body is at the point (− 4,0), moving towards the origin at 6 m s^{-1}, and losing speed at the rate of 20 m s^{-1} every second:

Fig. 1.1

The body's position is to the *left* of the origin, so $s = -4$ m.
The movement is to the *right*, so $v = +6$ m s^{-1}.
The *gain* of velocity is to the *left*, so $a = -20$ m s^{-2}.

Standard equations for motion with constant acceleration

Consider now a body which moves with a *constant* (or *uniform*) acceleration for a time t. The following five important quantities are associated with the motion.

Initial velocity, u
Final velocity, v
Time, t
Acceleration, a
Final displacement from starting point, s.

These five quantities are related by five standard equations, each of which omits one of them.

1. $v = u + at$
When the acceleration of a body is constant, its value is given by

$$a = \frac{\text{velocity gained}}{\text{time taken}}$$

Now the velocity gained is clearly the final velocity minus the initial velocity, i.e. $v - u$; hence

$$a = \frac{v - u}{t}$$

$$\therefore \quad at = v - u$$

$$\therefore \quad v = u + at.$$

2. $s = \frac{(u + v)t}{2}$
This equation follows from the fact that

$$\text{average velocity} = \frac{\text{increase in displacement}}{\text{time taken}}$$

For a body whose velocity is changing at a uniform rate, the average velocity is simply the average of the initial and final velocities. We therefore have

$$\frac{u + v}{2} = \frac{s}{t}$$

and hence

$$s = \frac{(u + v)t}{2}.$$

3. $s = ut + \frac{1}{2}at^2$
This and the remaining equations can be obtained by eliminating the appropriate quantity from equations 1 and 2. Here we eliminate v by substituting $u + at$ for v in equation 2.

$$s = \frac{(u + u + at)t}{2}$$

$$\therefore \quad 2s = 2ut + at^2$$

$$\therefore \quad s = ut + \frac{1}{2}at^2.$$

4. $s = vt - \frac{1}{2}at^2$

To obtain this equation we eliminate u from equations 1 and 2.

From equation 1 $\qquad\qquad\qquad\qquad u = v - at$

Substituting in equation 2 $\qquad\qquad s = \dfrac{(v - at + v)t}{2}$

$$\therefore\ 2s = 2vt - at^2$$

$$\therefore\ s = vt - \tfrac{1}{2}at^2.$$

5. $v^2 = u^2 + 2as$

The quantity t is not involved here, so we proceed as follows.

From equation 1 $\qquad\qquad\qquad\qquad t = \dfrac{v - u}{a}$

Substituting in equation 2 $\qquad s = \dfrac{(u + v)(v - u)}{2a}$

$$\therefore\ 2as = v^2 - u^2$$

$$\therefore\ v^2 = u^2 + 2as.$$

The complete set of equations is shown in the box below.

$$
\begin{array}{ll}
v = u + at & s = ut + \tfrac{1}{2}at^2 \\[2mm]
s = \dfrac{(u + v)t}{2} & s = vt - \tfrac{1}{2}at^2 \\[2mm]
& v^2 = u^2 + 2as
\end{array}
$$

Motion under gravity

An important example of motion with constant acceleration is that of a body moving vertically under the influence of gravity alone. Provided that the body stays reasonably close to the Earth's surface it moves with a constant downward acceleration which is denoted by g. The value of g varies slightly from place to place on the earth, being about $9.81\ \mathrm{m\,s}^{-2}$ in London and slightly less at the equator. In this book we shall always adopt the common practice of taking g to be $10\ \mathrm{m\,s}^{-2}$.

Worked examples (elementary)

The fact that each equation omits one quantity gives us a simple method of deciding which equation to use in problems. We tabulate all the information both given and required, which in elementary examples normally involves *four* of the five quantities. We then use the equation which omits the quantity not involved.

Example 1   A uniformly accelerating body starts with a velocity of 36 km h^{-1} and 5 s later has a velocity of 108 km h^{-1}. Find (a) the acceleration in m s^{-2}, (b) the distance travelled in the third second.

We begin by converting km h^{-1} to m s^{-1}. This is best done in two steps, as follows.

$$1 \text{ km h}^{-1} = 1000 \text{ m h}^{-1}$$

$$= \frac{1000}{60 \times 60} \text{ m s}^{-1}$$

$$= \tfrac{5}{18} \text{ m s}^{-1}.$$

(This result, namely $1 \text{ km h}^{-1} = \tfrac{5}{18} \text{ m s}^{-1}$, is worth learning by heart.) It follows that $36 \text{ km h}^{-1} = 10 \text{ m s}^{-1}$ and $108 \text{ km h}^{-1} = 30 \text{ m s}^{-1}$.

(a) We have $u = 10$, $v = 30$, $t = 5$, $a = ?$

Using

$$v = u + at,$$

we have

$$30 = 10 + 5a$$

from which

$$a = 4 \text{ m s}^{-2}.$$

(b) Since the motion is always in the same direction, the distance moved in the third second is the value of s when $t = 3$ (s_1 say) minus the value of s when $t = 2$ (s_2 say).

Using

$$s = ut + \tfrac{1}{2}at^2,$$

we have

$$s_1 = (10 \times 3) + (\tfrac{1}{2} \times 4 \times 9) = 48$$

and

$$s_2 = (10 \times 2) + (\tfrac{1}{2} \times 4 \times 4) = 28.$$

The distance travelled in the third second is thus **20 m**.

Example 2   A body accelerates uniformly at 8 m s^{-2} for 3 s, finishing with a velocity of 40 m s^{-1}. Find the distance it travels.

We have

$$a = 8, \ t = 3, \ v = 40, \ s = ?$$

Using

$$s = vt - \tfrac{1}{2}at^2,$$

we obtain

$$s = (40 \times 3) - (\tfrac{1}{2} \times 8 \times 9)$$

$$= 84 \text{ m}.$$

Example 3   A body thrown vertically upwards reaches a maximum height of 3.2 m. Find (a) its initial speed, (b) the total time the body takes to return to its starting point, (c) the total time it takes to reach a point 4 m below its starting point.

(a) At its maximum height the body has a velocity of zero. Hence for the upward journey, taking the upward direction as positive, we have

$$v = 0, \ a = -10, \ s = 3.2, \ u = ?$$

Using

$$v^2 = u^2 + 2as,$$

we obtain

$$0 = u^2 - (2 \times 10 \times 3.2),$$

from which

$$u = 8 \text{ m s}^{-1}.$$

(b) When the body has returned to its starting point $s = 0$. Hence, taking the period between the body's being thrown up and its returning, we have

$$u = 8, \ a = -10, \ s = 0, \ t = ?$$

Using

$$s = ut + \tfrac{1}{2}at^2,$$

we have

$$0 = 8t - 5t^2$$
$$= t(8 - 5t)$$
$$\therefore \ t = 0 \quad \text{or} \quad \tfrac{8}{5}.$$

Clearly $t = 0$ at the start of the motion and $\tfrac{8}{5}$ or 1.6 at the end; hence the total time taken is **1.6 s**.

(c) Taking the complete period between the start of the motion and the body's reaching a point 4 m below the starting point, and again letting the upward direction be positive, we have

$$s = -4, \ a = -10, \ u = 8, \ t = ?$$

Using

$$s = ut + \tfrac{1}{2}at^2,$$

we have

$$-4 = 8t - 5t^2$$
$$\therefore \quad 5t^2 - 8t - 4 = 0$$
$$\therefore \quad (5t + 2)(t - 2) = 0$$
$$\therefore \quad t = -\tfrac{2}{5} \quad \text{or} \quad 2.$$

Clearly we require the positive answer; hence the required time is **2 s**.

Example 4 A uniformly accelerating body travels 6 m and 14 m, respectively, in two successive seconds. Find the acceleration.

The best method here is to apply the equation $s = ut + \tfrac{1}{2}at^2$ to the first second and to the *complete period* of 2 s. The initial velocity u then is the same in both cases, and we obtain two simultaneous equations in u and a.

Taking the first second, we have

$$6 = (u \times 1) + (\tfrac{1}{2}a \times 1^2)$$

that is,

$$12 = 2u + a \quad (1)$$

In the complete period of 2 s

$$s = 6 + 14 = 20,$$

and we have

$$20 = (u \times 2) + (\tfrac{1}{2}a \times 2^2)$$

that is,

$$20 = 2u + 2a \quad (2)$$

Now we eliminate u by subtracting (1) from (2), obtaining $a = 8 \text{ m s}^{-2}$.

Exercise 1a

√1   A body starts at 10 m s^{-1}, and accelerates at 4 m s^{-2} for 3 s. Find the final speed and the distance travelled.

2   A body starts at rest, accelerates uniformly for 4 s, and finishes with a velocity of 20 m s^{-1}. Find the acceleration and the distance travelled.

√3   A body starts with a velocity of 9 km h^{-1}, and 2 s later has a velocity of 45 km h^{-1}. Find the acceleration in m s^{-2} and the distance travelled.

4   After 5 s a body accelerating at 6 m s^{-2} has a velocity of 180 km h^{-1}. Find the distance travelled and the initial velocity.

√5   A stone is dropped from rest at a point 20 m above the ground. Find the time it takes to reach the ground and its speed at the moment of impact.

6   A stone dropped from rest takes 0.5 s to reach the ground. Find the height from which it is dropped and its speed at the moment of impact.

√7   A body decelerates uniformly to rest from a speed of 72 km h^{-1} in a distance of 40 m. Find the deceleration and the time taken.

√8   A car travelling at 12 m s^{-1} decelerates at 2 m s^{-2}. How far does it travel before coming to rest?

√9   A ball is thrown vertically upwards with a speed of 8 m s^{-1}. Find (a) the time it takes to return to its starting point, (b) its distance below the starting point 2 s after being thrown.

√10   A body is thrown vertically upwards with a speed of 36 km h^{-1}. How long does it take to reach a point 15 m below the starting point?

11   A body accelerates at 5 m s^{-2} for 3 s, travelling 27 m. Find the velocities with which it starts and finishes.

√12   A ball is dropped from rest at the top of a building, and strikes the ground at 72 km h^{-1}. Find the height of the building.

13   A body thrown vertically upwards reaches a point 8 m below its starting point 2 s later. Find the speed with which it is thrown.

What does it mean exactly?   — 14   A body is thrown vertically upwards at 8 m s^{-1}. Find the two times at which it is 3 m above its starting point.

15   A body is thrown vertically upwards at 9 m s^{-1}. Find the interval of time for which it is more than 4 m above its starting point.

√16   A body thrown vertically upwards reaches a maximum height of 5 m. Find the speed with which it is thrown and the total time it is in the air.

hard! - √17   A uniformly accelerating body which starts from rest travels 2 m in the first second. How far does it travel in the third second?

???? —18   A uniformly accelerating body which starts from rest travels 36 m in the 5th second. Find the acceleration and the distance travelled in the 6th second.

√19   A uniformly decelerating train of length 40 m takes 6 s to pass completely through a station of length 50 m. Given that the train enters the station at 72 km h^{-1}, find its deceleration in m s^{-2}.

√20   A ball is thrown upwards to a height of 125 cm. For how long is it in the air?

21   A ball is thrown vertically upwards. Find its maximum height if it takes 0.6 s to reach it.

hard! √22   The initial speed of a uniformly accelerating body is 4 m s^{-1}, and after it has travelled 5 m its speed is 6 m s^{-1}. How long does it take to travel another 16 m?

???? —23   In two successive seconds a uniformly accelerating body travels 3 m and 7 m, respectively. Find its acceleration.

How do you prove it?

hard

24 A uniformly accelerating body travels 3 m and 9 m, respectively, in its first two seconds. How far does it travel in its fifth second?

25 A stone is thrown vertically upwards from the edge of a cliff 25 m high at 20 m s^{-1}. Find the time it takes to reach the water at the foot of the cliff.

26 A stone is thrown down from the top of a cliff 35 m high and enters the water at the foot with a speed of 40 m s^{-1}. Find the speed at which it is thrown.

27 A uniformly decelerating body covers successive 100 m distances in 5 s and 10 s. Find the initial speed, the deceleration, and the further time required for the body to come to rest.

28 A uniformly decelerating train of length 20 m enters a station of length 40 m. The front of the engine leaves the station 5 s later, and the end of the train leaves the station after a further 5 s. Find the deceleration of the train.

29 A ball thrown straight up into the air reaches its greatest height of H in a time of T. Prove that $2H = gT^2$.

30 A uniformly accelerating body starts with a speed of u, and in successive times of t travels distances of s and $2s$. Prove that its acceleration is $4u^2/s$.

More complicated examples

There are two main ways in which complications can be introduced: the motion can be in several stages (e.g. an acceleration followed by a deceleration), or more than one moving body can be involved.

Velocity–time graphs

When the motion of a body is in several stages it is often helpful to consider a graph of velocity against time. For example, Fig. 1.2 illustrates the motion of a body which accelerates from rest, moves with constant velocity for a time, then decelerates back to rest.

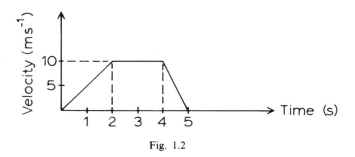

Fig. 1.2

Velocity–time graphs have the following important properties.

> **Gradient = acceleration.**
> **Area between graph and time-axis = distance travelled.**

Thus in the above example the acceleration is $10/2 = 5\,\mathrm{m\,s^{-2}}$, the deceleration is $10\,\mathrm{m\,s^{-2}}$, and the distance travelled in the first 2 s is $\frac{1}{2} \times 2 \times 10 = 10$ m.

Worked examples (more complicated)

Example 1   A body accelerates uniformly from rest to a speed of $10\,\mathrm{m\,s^{-1}}$, then moves at constant speed for a time, and finally decelerates uniformly to rest. If the acceleration is twice the deceleration, and the total distance and total time are 120 m and 18 s, find the acceleration.

The velocity–time graph is shown in Fig. 1.3. Letting $OE = t$, we have $DC = 2t$, since the gradient of OA is numerically twice that of BC.

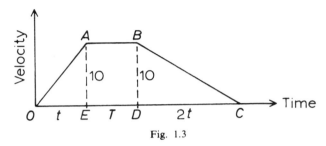

Fig. 1.3

The total distance is simply the area below the graph. Hence we have:

$$\text{Total distance} = OAE + ABDE + BCD.$$

That is,

$$120 = 5t + 10T + 10t$$

$$= 15t + 10T$$

$$\therefore\ 3t + 2T = 24 \quad (1)$$

Also, since total time = 18 s, we have

$$3t + T = 18. \quad (2)$$

Subtracting (2) from (1) we now obtain $T = 6$, and substituting this value into (2) we get $t = 4$.

$$\text{Hence acceleration} = AE/OE = 10/4 = \mathbf{2.5\,m\,s^{-2}}.$$

Example 2   A body P starts with a speed of $2\,\mathrm{m\,s^{-1}}$, and moves along a straight line with a constant acceleration of $4\,\mathrm{m\,s^{-2}}$. Three seconds later another body Q starts from the same point with a speed of $8\,\mathrm{m\,s^{-1}}$, and moves along the same straight line with an acceleration of $10\,\mathrm{m\,s^{-2}}$. Find the distance from the starting point at which the bodies are together.

Let the time taken by P to reach the meeting point be t; then the time taken by Q will be $t - 3$. Let s be the required distance.
   Applying $s = ut + \frac{1}{2}at^2$ to each body in turn, we have

$$s = 2t + 2t^2$$

and

$$s = 8(t - 3) + 5(t - 3)^2.$$

Eliminating s:

$$2t + 2t^2 = 8t - 24 + 5t^2 - 30t + 45.$$

$$\therefore \quad 3t^2 - 24t + 21 = 0$$

$$\therefore \quad (3t - 3)(t - 7) = 0$$

$$\therefore \quad t = 1 \quad \text{or} \quad 7.$$

Since the second body Q does not start its journey until the time is 3 seconds, $t = 1$ is impossible. Hence $t = 7$. We find the required distance by substituting this value into the first of the above two equations for s:

$$s = 2t + 2t^2$$

$$= 14 + 98 = 112.$$

The distance travelled by both bodies when they meet is thus **112 m**.

Exercise 1b ~~for All use graphical (just for practice)~~

1 A body starts from rest, accelerates uniformly at $4 \, \text{m s}^{-2}$ for 2 s, then decelerates uniformly to rest. If the total distance travelled is 20 m, what is the deceleration?

2 A body starts at rest, travels 8 m at a constant acceleration, then 16 m at constant speed, and finally 4 m at a constant deceleration, finishing at rest. Given that the total time taken is 10 s, find the acceleration and the deceleration.

3 Two bodies start together and move along the same straight line. If one moves with a constant speed of $6 \, \text{m s}^{-1}$, while the other starts from rest and accelerates at $4 \, \text{m s}^{-2}$, how long will it be before they are together again?

4 A man starts walking at a steady speed of $1 \, \text{m s}^{-1}$, and 6 s later his son sets off from the same point in pursuit of him, starting at rest and accelerating at $2/3 \, \text{m s}^{-2}$. How far do they go before they are together?

5 A body starts at rest, accelerates uniformly for a time, then decelerates to rest at twice the rate. The total distance and total time are 225 m and 15 s. Find the acceleration.

6 A body travels at a constant speed for 5 s, accelerates at a constant rate for the next 2 s, then travels at a constant speed again for another 5 s. Given that it travels 120 m altogether, and the final speed is four times the initial speed, find the initial speed and the acceleration.

7 One body is dropped from rest at a height of 10 m, and at the same time another body is thrown upwards from the ground. If they cross at a height of 5 m, what is the initial speed of the second body?

8 A boy runs at $4 \, \text{m s}^{-1}$ away from a cyclist who starts at rest and accelerates at $2 \, \text{m s}^{-2}$. If the boy has an initial lead of 5 m, how long does the cyclist take to catch him?

9 One body is thrown vertically upwards with an initial speed of $20 \, \text{m s}^{-1}$, and 2 s later another body is thrown vertically upwards from the same place with the same speed. At what height will they collide?

10 A body starts with a velocity of $20 \, \text{m s}^{-1}$, accelerates uniformly for a time t, moves with constant velocity for the same time, and finally decelerates to rest, again taking a time t. Given that the final deceleration is three times the original acceleration, and the total distance travelled is 140 m, find t.

11   Bodies A and B start together and move along the same straight line. A starts with a speed of 10 m s^{-1} and moves with a constant deceleration, while B starts at 5 m s^{-1} and accelerates at 4 m s^{-2}. Find the deceleration of A if they meet when the velocity of B is twice that of A.

12   Bodies A and B start 210 m apart. A starts to move towards B with an initial speed of 4 m s^{-1} and a steady acceleration of 2 m s^{-2}, and 2 s later B sets off towards A at a constant speed of 5 m s^{-1}. Find the distance travelled by A when they meet.

13   A body starts with a speed of u, decelerates uniformly to half this speed while travelling a distance s, travels the same distance at constant speed, then decelerates to rest at twice its original rate. Prove that the total time taken is $4s/u$.

14   A body starts at rest, moves with a constant acceleration of a for a time t, then moves with constant speed for time T, and finally decelerates to rest at a quarter of its original acceleration. Given that the average speed for the whole journey (total distance/total time) is $4at/5$, prove that $T = 15t/2$.

2

Newton's Laws of Motion. The Equation $F = ma$

Before stating Newton's laws we must explain the terms *mass* and *momentum*. The mass of a body is a measure of the *quantity of matter* it contains. Mass is one of the three fundamental quantities in mechanics (the others being length and time) and its SI unit is the base unit the *kilogram*. Mass must not be confused with *weight*, which in everyday life is also measured (incorrectly) in kilograms, pounds, etc. and which is defined as the *force* on a body due to gravity. Weight is proportional to mass (at a given place on earth) but the two concepts are clearly distinct since a body taken outside the earth's gravitational field loses its weight but retains its mass.

The *linear momentum of a body* (usually called simply its *momentum* – there is another quantity called *angular momentum* which will not be considered in this book) is a measure of the difficulty that is experienced in stopping it. Momentum is defined as *mass multiplied by velocity*.

Newton's laws may now be stated as follows:

Law 1   *A body which has no resultant force on it either remains at rest or moves in a straight line at constant speed.*
Law 2   *The magnitude of a force is proportional to the rate of change of momentum that it causes.*
Law 3   *Action and reaction are equal and opposite.* (That is, if a body A is exerting a force on a body B, then B must be exerting an equal and opposite force on A.)

The equation $F = ma$ and the SI unit of force

Newton's second law may be stated mathematically in the following way.

$$F = k \frac{\mathrm{d}}{\mathrm{d}t} (mv).$$

This equation can be simplified. Bodies are normally of constant mass (some exceptions are snowballs and raindrops) and provided that this is so $\mathrm{d}(mv)/\mathrm{d}t$ is equal to $m \, \mathrm{d}v/\mathrm{d}t$. Also, in the SI system of units, the constant of proportionality k is taken to be 1 when m is in kg and v is in m s^{-1}. We thus have

$$F = m \frac{\mathrm{d}v}{\mathrm{d}t}$$

or

$$F = ma$$

From this equation it can be seen that the SI unit of force is the kg m s^{-2}, which is given the name *newton* and denoted by N. We can therefore define the newton as follows.

> **The newton is the force that causes an acceleration of $1\,\text{m s}^{-2}$ when acting on a body of mass 1 kg.**

Weight

As stated above, the definition of weight is *the force on a body due to gravity*. It must be stressed that weight is the force *on* a body and not the force which the body exerts on something else, such as support. The force that a body applies to something supporting it may or may not be equal to its own weight. (See worked examples below.)

Since $F = ma$ and the acceleration due to gravity is the constant g (which is the same for all bodies at a given place on earth), we clearly have the following relationship between a body's weight W and its mass m.

$$W = mg$$

Thus, for example, taking g to be $10\ \text{m s}^{-2}$, we can immediately obtain the weight of any body in newtons by multiplying its mass in kilograms by 10. For example,

$$\text{weight of a 2 kg body} = 20\ \text{N},$$
$$\text{weight of a 500 g body} = \ \ 5\ \text{N},$$
$$\text{etc.}$$

Forces between bodies in contact

For genuine contact to occur between two bodies, each must exert a force on the other which is perpendicular to their common surface. The two forces are equal, by Newton's third law, and both are usually called the *normal reaction* and denoted by R. If it is necessary to show both the forces in a diagram, the diagram should be drawn with the two bodies slightly apart (Fig. 2.1).

Fig. 2.1

If, in addition to the normal reaction, there are forces between the bodies acting *along* the surface of contact, these must be due to *friction*, which only comes into play in response to some attempt to slide one body along the surface of the other.

We shall consider friction in detail in a later chapter. Bodies described as 'smooth' (i.e. frictionless) can only have the normal reaction acting between them.

Tensions in strings

Consider a string which is subjected to a pull at each end. By Newton's third law such a string will react by pulling back with an equal force on each of the objects applying the pull. This force is called the *tension* in the string (Fig. 2.2).

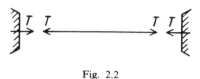

Fig. 2.2

Figure 2.2 shows the forces *on* the string, and also those exerted *by* the string. In practice the simplified type of diagram shown in Fig. 2.3 is used, the convention being that arrows on strings denote the forces exerted *by* the strings.

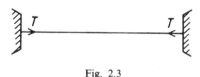

Fig. 2.3

If the string is stationary or moving at constant speed, the tensions at the two ends must clearly be equal since the string itself is in equilibrium. Even if the string is accelerating, however, the tensions at the ends are in practice considered to be equal, provided that the string is subjected to no forces along its length other than those at the ends. This is because strings are normally regarded as 'light' or 'weightless', and thus as having zero mass. There can be no resultant force on a body of this kind (as there would be if the tensions at the ends were different) since such a force would produce an infinite acceleration.

At any point in a string an imaginary division can be made and we can consider the forces between the two parts. It follows from the reasoning just given that these forces must also equal the tensions at the ends, and thus that we can speak of there being the same tension throughout the entire length of a (weightless) string.

Worked examples

Note: As explained at the beginning of the book, a useful convention in diagrams involving both forces and accelerations is to use a single arrowhead ($\longrightarrow$) to represent forces, and a double arrowhead ($\longrightarrow\!\!\!\!>$) to represent accelerations.

Example 1   A 500 g block is pushed along a rough horizontal surface with an acceleration of 4 m s^{-2}. If the frictional force is half the weight of the block, what is the applied force (Fig. 2.4)?

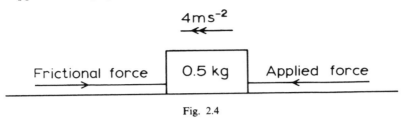

Fig. 2.4

The information given enables us to find the *resultant* force (the net effect of the two forces acting) by using $F = ma$:

$$\text{resultant force} = 0.5 \times 4 = 2 \text{ N}.$$

Also the weight of the block is given by $W = mg$; thus the weight is $0.5 \times 10 = 5$ N and hence the frictional force is 2.5 N.
Now since

$$\text{resultant force} = \text{applied force} - \text{frictional force},$$

we have

$$\text{applied force} = \text{resultant force} + \text{frictional force}$$

$$= 2 + 2.5$$

$$= \textbf{4.5 N}.$$

Example 2   A force of 70 N acts vertically upwards on a 5 kg body. Find the acceleration.

The weight of a 5 kg body is $5g = 50$ N; hence the forces on the body are as shown in Fig. 2.5. Clearly the resultant upward force is 20 N. Using $F = ma$, we thus have

$$a = F/m = 20/5 = \textbf{4 m s}^{-2}.$$

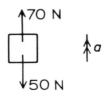

Fig. 2.5

Example 3   A force of $5W$ acts vertically upwards on a body of weight $3W$. Find the acceleration in terms of g.

The resultant upward force is $5W - 3W = 2W$. From the general equation $W = mg$, which relates weight and mass for all bodies, it follows that the mass of a body of weight $3W$ is $3W/g$. Hence, using $F = ma$, we have

$$a = F/m = 2W \times g/3W = 2g/3.$$

Example 4 A body of mass $4M$ stands on a platform which is accelerating downwards at $0.25\,g$. Find the reaction between the body and the platform (Fig. 2.6).

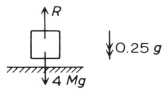

Fig. 2.6

From the equation $F = ma$, the resultant downward force on the body is $4M \times 0.25\,g = Mg$. Thus since

$$\text{resultant downward force} = \text{weight} - \text{reaction},$$

we have

$$Mg = 4\,Mg - R$$

and hence

$$R = 3Mg.$$

Note: The reaction just found is the same as the *apparent weight* of the body when it is accelerating on a platform as described. A weighing machine simply measures the force that is applied to it, that is the reaction between itself and the body it is weighing.

Example 5 A pan of mass 100 g carrying a body of mass 300 g is lifted vertically upwards by a force of 6 N. Find the reaction between the pan and the weight.

Fig. 2.7   Fig. 2.8

In Fig. 2.7 the pan and weight are being considered together, as a single entity, and only the *external* forces on this entity are shown. (The two *internal* forces, which the pan and weight exert on each other, cancel each other out.) Considering the whole system we have

$$\text{resultant upward force} = 6 - 4 = 2\,\text{N},$$
$$\text{total mass} = 400\,\text{g} = 0.4\,\text{kg};$$
$$\text{hence } a = F/m = 2/0.4 = 5\,\text{m s}^{-2}.$$

Next we consider the weight alone (Fig. 2.8). (The pan alone would also be possible.) There are just two forces on this, namely its own weight and the reaction R provided by the pan. Applying $F = ma$ and remembering that the acceleration is upwards, we have

$$R - 3 = 0.3 \times 5$$

from which

$$R = 4.5\,\text{N}.$$

Example 6   Three coupled trucks, with masses 50, 60 and 90 tonnes, are pulled along a horizontal track by a locomotive which applies a force of 50 kN to the first truck. Find the tensions in the couplings (Fig. 2.9).

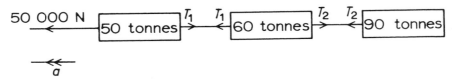

Fig. 2.9

Consider the whole system. Applying $F = ma$, we have

$$a = \frac{50\,000}{200\,000} = 0.25 \text{ m s}^{-2}.$$

Now to determine the tensions we apply $F = ma$ to convenient parts of the system. To find T_1 the best part to take is the 60 and 90 tonne trucks together. The two forces of T_2 then become internal forces which cancel each other out, and we have simply

$$T_1 = 150\,000 \times 0.25$$
$$= 37\,500 \text{ N or } \textbf{37.5 kN}.$$

To find T_2 we apply $F = ma$ to the 90 tonne truck alone:

$$T_2 = 90\,000 \times 0.25$$
$$= 22\,500 \text{ N or } \textbf{22.5 kN}.$$

Example 7   A jet of water of cross-sectional area 20 cm^2 and speed 4 m s^{-1} strikes a vertical wall. Assuming that no rebounding takes place, find the force that the jet exerts on the wall. (Take the density of water to be 1000 kg m^{-3}.)

This example illustrates the one elementary type of question which requires the use of the strict equation *Force = rate of change of momentum* rather than the simple $F = ma$. Since the jet exerts a continuous force on the wall, the latter must exert an equal reactive force on the water (by Newton's third law) and the effect of this force is to destroy the water's momentum. The magnitude of the force is simply the amount of momentum destroyed per second. We proceed as follows:

$$\text{Cross-section} = 20 \text{ cm}^2 = (20 \times 100^{-2})\text{ m}^2 = 0.002 \text{ m}^2.$$

Now volume of water striking wall per second = cross-section × speed

$$= 0.002 \times 4 = 0.008 \text{ m}^3.$$

Hence mass of water striking wall per second = volume × density

$$= 0.008 \times 10^3 = 8 \text{ kg}.$$

All of this mass is reduced from a speed of 4 m s^{-1} to rest; hence the momentum destroyed per second is 32 kg m s^{-1} and the force on the wall is **32 N**.

Exercise 2

1 State the weight in N of (a) a 65 kg body, (b) a 35 kg body.
2 What are the masses of bodies weighing (a) 45 N, (b) 0.2 N?
3 A horizontal force of 40 N acts for 3 s on a 20 kg body, initially at rest. Find the acceleration and the final speed.
4 What constant force is needed to move a 2 kg body, initially at rest, a distance of 500 cm in 2 s?
5 A force of 2 N, acting on a certain body, increases its speed from 3 km h^{-1} to 11 km h^{-1} in 2 s. Find the acceleration in m s^{-2} and the mass of the body.
6 Find the constant retarding force which will bring a 2 kg body from a speed of 3 m s^{-1} to rest in 5 s.
7 What force will make a body of weight 5 N accelerate uniformly from rest to a speed of 36 km h^{-1} in a distance of 2.5 m?
8 A 5 tonne body is subjected to a force of 20 000 N for 3 s. Find its gain of speed.
9 A horizontal force of 3W acts on a body of weight 4W. Find its acceleration in terms of g.
10 A force of 200 N acts vertically upwards on a 15 kg body. Find the acceleration.
11 An upward force of 45 000 N acts on a 3 tonne body. Find the acceleration.
12 A force of 7W acts vertically upwards on a body of weight 3W. Find the acceleration.
13 What force, acting vertically upwards on a 250 g body, will cause it to accelerate (a) upwards at 3 m s^{-2}, (b) downwards at 6 m s^{-2}?
14 A force of 15 N acts vertically upwards on a body of weight 20 N. Find the downward acceleration.
15 An upward force of 2.5 mg acts on a body of mass 2 m. Find its acceleration in terms of g.
16 An upward force of x acts on a body of mass y. Find the acceleration in terms of x, y and g.
17 An upward force of 25 N, acting on certain body, causes it to accelerate upwards at 5 m s^{-2}. Find the mass of the body.
18 With an upward force of 40 N acting on it, a certain body accelerates downwards at 2 m s^{-2}. Find the mass of the body.
19 A 500 g block is pushed along a horizontal table against a frictional force of 6 N. If the acceleration is 8 m s^{-2}, what is the applied force?
20 A 2 kg block is pushed along a rough horizontal table by a force of 12 N. If it takes 1.5 s to accelerate from 1 m s^{-1} to 7 m s^{-1}, what is the frictional force?
21 A 30 kg block stands on the floor of a lift. Find the force between the block and the floor if the lift accelerates (a) downwards at 0.5 g, (b) upwards at 2 g, (c) downwards at g.
22 A 75 kg man stands on a weighing machine in a lift which accelerates (a) upwards, (b) downwards, at 0.2 g. What weights are recorded?
23 What force, acting vertically upwards on a 20 kg body, will accelerate it from rest to a speed of 72 km h^{-1} in 2 s?
24 A 2 tonne body, falling from rest against a constant resistance, travels 16 m in the first 4 s. Find the resistance.
25 A force of 4W, acting vertically downwards on a certain body, causes it to accelerate at 3 g. Find the mass of the body in terms of W and g.

26   A pan of mass 50 g carrying a body of mass 250 g is lifted vertically upwards by a force of 4.5 N. Find (a) the acceleration, (b) the reaction between the pan and the weight.

27   A 5 kg plank carrying a 10 kg block is lifted vertically upwards by a force of 200 N. Find the reaction between the plank and the block.

28   A pan of mass 100 g carries a body of mass 200 g. A person holds the pan and lifts the system with an acceleration of 5 m s^{-2}. Find (a) the force applied by the person to the pan, (b) the reaction between the pan and the weight.

29   A block of weight $2W$ stands on a block of weight $3W$, and an upward force of $20W$ is applied to the latter. Find the reaction between the blocks.

30   A man holds a bag of mass 500 g which contains a body of mass $2\frac{1}{2}$ kg. If he lowers it with an acceleration of 4 m s^{-2}, what are the forces (a) applied by the man to the bag, (b) applied by the bag to the body it contains?

31   A pile of blocks with masses (from the bottom up) of 4 kg, 3 kg, 2 kg and 1 kg is raised by a person holding the bottom block. When the force between the middle two blocks is 105 N, what are the forces (a) between the top two blocks, (b) between the person and the bottom block?

32   A platform of mass 100 tonnes carries a body of mass 50 tonnes and the system is lowered with a certain acceleration. If the reaction between the two is 300 kN, what is the resultant force on the platform?

33   Blocks with masses of 200 g, 300 g, 400 g and 600 g are connected by strings and the system is pulled along a smooth table by a force of 3 N applied to the first block. Find the tensions in the strings.

34   Bodies of mass 5 kg, 7 kg and 8 kg are connected by strings and pulled along a smooth table by a force of 30 N applied to the first body. Find the tension in the strings.

35   Bodies of mass $4m$, $3m$ and $2m$ are connected by strings and pulled along a smooth table by a force applied to the first body. Given that the tension in the string between the $2m$ and $3m$ bodies is $4mg$ find (a) the applied force, (b) the tension in the other string, in terms of m and g.

36   A jet of water of cross-section 30 cm^2 and speed 5 m s^{-1} strikes a plane surface, all its momentum thereby being lost. Find the force exerted on the surface.

37   A jet of water of cross-section 20 cm^2 and speed 8 m s^{-1} strikes a plane surface and rebounds at a speed of 2 m s^{-1}. Find the force on the surface.

38   A pipe of cross-section 10 cm^2 delivers 8000 cm^3 of water per second in the form of a jet which strikes a wall. If no rebounding occurs, find the force on the wall.

39   A force of 72 N is provided by a jet of water which strikes a wall at 6 m s^{-1} and is thereby brought to rest. Find the cross-sectional area of the jet.

40   A force of 20 N is provided by a jet of water of cross-section 8 cm^2 which strikes a wall and does not rebound. Find the speed of the jet.

41   A pipe of cross-section 50 cm^2 delivers 0.12 m^3 of water per minute in the form of a jet which strikes a plane surface and is thereby brought to rest. Find the force on the surface.

42   A force of 80 N is provided by a jet of water which is brought to rest from a speed of 4 m s^{-1}. Find the cross-sectional area of the jet.

43   A force of 60 N is provided by a jet of water of cross-section 10 cm^2 which strikes a wall and does not rebound. Find the speed of the jet.

3

Vectors, Resolution, Components

In this chapter we give only the vector theory needed for an understanding of the rest of the book, in which the *language* of vectors is often used, but advanced vector theory is not required.

Definitions and notation

Vectors may be defined as **quantities possessing magnitude and direction which can be combined by application of the triangle or parallelogram laws** (see below). They are to be contrasted with *scalars*, which are quantities possessing magnitude only. Examples of vectors are *displacement*, *velocity*, *momentum*, *acceleration* and *force*.

Vectors are usually represented by directed line-segments, as shown in Fig. 3.1. This diagram represents a vector which can be denoted either by $\vec{AB}$ or by the single small letter **a**. The latter is either printed in bold type or underlined. The magnitude of the vector is represented by the length of the line-segment, and it may be denoted by AB, a, or $|\mathbf{a}|$.

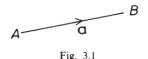

Fig. 3.1

Free vectors are vectors whose values are not considered to be affected by changes in position; they are completely specified by their magnitudes and directions. *Localised* vectors, on the other hand, are affected in value by changes in position. *Forces*, for example, are localised vectors because a change in the line of action of a force alters its effect upon a body and thus its value.

The *negative* of a vector $\vec{AB}$, denoted by $-\vec{AB}$, is defined to be a vector which is equal in magnitude and parallel to $\vec{AB}$, but opposite to it in direction or 'sense'.

The vector $n\mathbf{a}$ (i.e. the product of a vector with a scalar) is defined to be a vector with magnitude na which has the same direction as **a** if n is positive and the opposite direction if n is negative.

The triangle and parallelogram laws

These two laws are equivalent, and they both define the *sum* or *resultant* of two vectors. The laws are illustrated by Figs. 3.2 and 3.3 respectively.

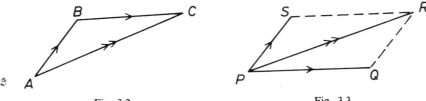

Fig. 3.2   Fig. 3.3

With reference to these two diagrams the triangle law states that

$$\overrightarrow{AB} + \overrightarrow{BC} = \overrightarrow{AC}$$

and the parallelogram law states that

$$\overrightarrow{PQ} + \overrightarrow{PS} = \overrightarrow{PR}.$$

For any physical quantity, such as force, to count as a vector, it must obey these laws.

As aids to remembering the laws correctly, note the following points.

1 *In the triangle law, we add the two 'clockwise' vectors to obtain the 'anti-clockwise' vector, or vice versa.*
2 *In the parallelogram law the three line-segments meet at a point, and all three are directed away from that point.*

Subtraction

The triangle and parallelogram laws can be used to subtract as well as add vectors, since subtraction is defined as the *addition of the negative*. Thus to subtract the vector **a**, we add the vector − **a**.

Resolution of a vector; rectangular components

We can re-express a vector as two separate vectors in *any* two (coplanar) directions by using the converse of the parallelogram law. This is called *resolving* the vector into two *components*. Normally, however, vectors are resolved into *rectangular* components – that is, components at right angles to each other – mainly because the values of these are easily calculable by trigonometry.

Example Suppose we wish to resolve a vector of magnitude 10 into two components making angles of 40° and 50° with it.

The parallelogram of vectors is now a rectangle, as shown in Fig. 3.4.

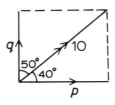

Fig. 3.4

By elementary trigonometry $p = 10 \cos 40°$ and $q = 10 \cos 50°$. Clearly therefore we have the following general rule for obtaining rectangular components.

Component = magnitude of vector × cosine of angle between vector and component.

In this book, we shall normally use the term 'component' to mean 'rectangular component'.

The unit vectors i, j

With reference to a pair of x and y axes, we can express any vector in the plane of the paper in terms of the two vectors **i** and **j**, each 1 unit in magnitude, whose directions are shown in Fig. 3.5.

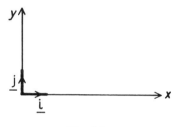

Fig. 3.5

(To deal with vectors in all directions a third unit vector **k**, perpendicular to the plane of **i** and **j**, is introduced. In this book, however, the treatment is limited to two dimensions.)

The vectors **i** and **j** are free – they need not pass through the origin – but their magnitudes and directions are fixed and they are therefore constant and not variable vectors.

Any vector in the x–y plane can now be expressed in the form $a\mathbf{i} + b\mathbf{j}$, the numbers a and b being the magnitudes of the x and y components of the vector. For example, the vectors $\mathbf{p} = 2\mathbf{i} - 5\mathbf{j}$ and $\mathbf{q} = -\mathbf{i} - \mathbf{j}$ may be represented as shown in Fig. 3.6.

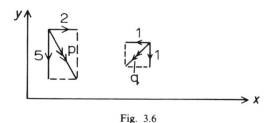

Fig. 3.6

One of the advantages of the representation in terms of **i** and **j** is that it makes addition and subtraction very simple. Since it can easily be shown that we can add vectors by adding their components, the sum of a set of vectors is obtained simply by adding all the coefficients of **i** and all those of **j**. For example,

$$(4\mathbf{i} - 2\mathbf{j}) + (6\mathbf{i} + 7\mathbf{j}) = 10\mathbf{i} + 5\mathbf{j}.$$

Both the magnitude and direction of a vector can be very easily obtained from its representation in terms of **i** and **j**. Consider once more the vector $a\mathbf{i} + b\mathbf{j}$, and let the angle it makes with the positive x-axis be θ, as shown in Fig. 3.7.

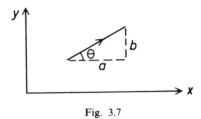

Fig. 3.7

Clearly we have

$$|a\mathbf{i} + b\mathbf{j}| = \sqrt{a^2 + b^2}$$
$$\text{and} \quad \tan\theta = b/a$$

Thus for example the magnitude of the vector $-4\mathbf{i} + 3\mathbf{j}$ is 5, and the angle it makes with the positive x-axis is arc tan $(-\frac{3}{4})$. If we require the unit vector in the direction of the vector given, we simply divide the latter by its magnitude, obtaining $(-4\mathbf{i} + 3\mathbf{j})/5$.

Position vectors

A position vector is a displacement vector by means of which the position of one point is specified relative to another. Thus for example the vector $\overrightarrow{OA}$ specifies the position of the point A relative to that of O. Position vectors expressed in terms of **i** and **j** are normally understood to specify points relative to the origin; so that, for example, the position vector $2\mathbf{i} - 3\mathbf{j}$ specifies the point $(2, -3)$.

A useful theorem for position vectors

Let the position vectors relative to O of the points A and B be **a** and **b**:

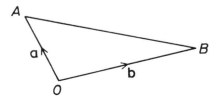

Fig. 3.8

Since, by the triangle law, $\mathbf{a} + \overrightarrow{AB} = \mathbf{b}$, it follows that

$$\overrightarrow{AB} = \mathbf{b} - \mathbf{a}.$$

Example If the position vectors of the points P and Q are $\mathbf{p} = 2\mathbf{i} - 3\mathbf{j}$ and $\mathbf{q} = 5\mathbf{i} + \mathbf{j}$, we have

$$\overrightarrow{PQ} = \mathbf{q} - \mathbf{p} = 3\mathbf{i} + 4\mathbf{j}.$$

Example of a problem

Three forces act on a 4 kg particle, causing it to accelerate at $5\,\mathrm{m\,s}^{-2}$ in the direction of the negative *x*-axis. One of the forces has a magnitude of 20 N and is in the direction of the vector $4\mathbf{i} - 3\mathbf{j}$, while another has a magnitude of 30 N and is in the direction of the vector $3\mathbf{i} + 4\mathbf{j}$. Express the third force in terms of $\mathbf{i}$ and $\mathbf{j}$.

The acceleration can be expressed as $-5\mathbf{i}$, and since the mass of the particle is 4 kg, the resultant force on it can be expressed as $-20\mathbf{i}$.

The first of the three forces acting on the particle is in the direction of the vector $4\mathbf{i} - 3\mathbf{j}$. This vector has a magnitude of 5, while the magnitude of the force is 20 N. It follows that this force ($\mathbf{F}_1$, say) can be expressed in terms of $\mathbf{i}$ and $\mathbf{j}$ by multiplying $4\mathbf{i} - 3\mathbf{j}$ by 4:

$$\mathbf{F}_1 = 16\mathbf{i} - 12\mathbf{j}.$$

By similar reasoning, since the second force is in the direction of the vector $3\mathbf{i} + 4\mathbf{j}$ and has a magnitude of 30 N, we can write

$$\mathbf{F}_2 = 6(3\mathbf{i} + 4\mathbf{j})$$
$$= 18\mathbf{i} + 24\mathbf{j}.$$

Now let the third force on the particle be $a\mathbf{i} + b\mathbf{j}$. Since the resultant force is $-20\mathbf{i}$, we have

$$16\mathbf{i} - 12\mathbf{j} + 18\mathbf{i} + 24\mathbf{j} + a\mathbf{i} + b\mathbf{j} = -20\mathbf{i},$$
$$\text{i.e.} \quad (34 + a)\mathbf{i} + (12 + b)\mathbf{j} = -20\mathbf{i}.$$

If follows that

$$34 + a = -20 \qquad\qquad 12 + b = 0$$
$$\therefore\ a = -54; \qquad\qquad \therefore\ b = -12.$$

The third force can therefore be expressed as $-54\mathbf{i} - 12\mathbf{j}$.

Exercise 3

1 $ABCDE$ is a pentagon. Express the following as single vectors: (a) $\overrightarrow{AE} + \overrightarrow{ED}$, (b) $\overrightarrow{CB} + \overrightarrow{EC}$, (c) $\overrightarrow{BD} - \overrightarrow{CD}$ (let $-\overrightarrow{CD} = \overrightarrow{DC}$), (d) $\overrightarrow{AB} - \overrightarrow{AD}$, (e) $\overrightarrow{BD} + \overrightarrow{AC} + DA$, (f) $\overrightarrow{AC} - \overrightarrow{AB} + \overrightarrow{CE}$.

2 $ABCD$ is a parallelogram whose diagonals meet at E. Letting $\overrightarrow{AB} = \mathbf{p}$ and $\overrightarrow{AD} = \mathbf{s}$, express the following vectors in terms of $\mathbf{p}$ and $\mathbf{s}$: (a) $\overrightarrow{AC}$, (b) $\overrightarrow{CB}$, (c) $\overrightarrow{BD}$, (d) $\overrightarrow{CE}$, (e) $\overrightarrow{EB}$.

3   *ABCDEF* is a regular hexagon. Letting $\overrightarrow{AB} = \mathbf{p}$ and $\overrightarrow{AF} = \mathbf{s}$, express the following vectors in terms of $\mathbf{p}$ and $\mathbf{s}$: (a) $\overrightarrow{CF}$, (b) $\overrightarrow{AD}$, (c) $\overrightarrow{EC}$, (d) $\overrightarrow{CA}$. (Divide the hexagon into equilateral triangles.)

4   *O* is the origin, and *A*, *B*, *C* are the points $(1, 4)$, $(-2, 3)$, $(-5, -1)$, respectively. Express the following vectors in terms of $\mathbf{i}$ and $\mathbf{j}$: (a) $\overrightarrow{OA}$, (b) $\overrightarrow{BO}$, (c) $\overrightarrow{CA}$, (d) $\overrightarrow{BC}$.

5   Find the magnitudes and inclinations to the *x*-axis (taking anticlockwise as positive) of the following vectors: (a) $4\mathbf{i} + 3\mathbf{j}$, (b) $-2\mathbf{i} + 2\mathbf{j}$, (c) $24\mathbf{i} - 10\mathbf{j}$, (d) $-6\mathbf{i} - 9\mathbf{j}$.

6   The points *A*, *B*, *C* have position vectors $3\mathbf{i} + \mathbf{j}$, $-2\mathbf{i} + 5\mathbf{j}$, $-4\mathbf{j}$, respectively. Express in terms of $\mathbf{i}$ and $\mathbf{j}$: (a) $\overrightarrow{AB}$, (b) $\overrightarrow{CB}$, (c) $2\overrightarrow{OA} - 3\overrightarrow{BO}$, (d) $3\overrightarrow{CA} + 4\overrightarrow{BA}$.

7   *OABC* is a parallelogram, *A* and *C* having position vectors $7\mathbf{i} + 3\mathbf{j}$ and $2\mathbf{i} + 9\mathbf{j}$, respectively. Find the position vectors of (a) *B*, (b) the mid-point of *AC*, (c) the point dividing *OB* in the ratio $1:2$.

8   The points *P*, *Q*, *R*, *S* have position vectors $6\mathbf{i} - \mathbf{j}$, $3\mathbf{i} + 7\mathbf{j}$, $-8\mathbf{i} + 11\mathbf{j}$, $-2\mathbf{i} - 5\mathbf{j}$, respectively. Express $\overrightarrow{PQ}$ and $\overrightarrow{SR}$ in terms of $\mathbf{i}$ and $\mathbf{j}$ and deduce the nature of the figure *PQRS*.

9   *A*, *B*, *C* are the points $(6, -8)$, $(1, 4)$, $(-3, 7)$, respectively. Express in terms of $\mathbf{i}$ and $\mathbf{j}$ (a) a vector of magnitude 10 in the direction of $\overrightarrow{CB}$, (b) a unit vector in the direction of $\overrightarrow{OA}$, (c) a vector of magnitude 65 in the direction of $\overrightarrow{AB}$.

In the remaining questions, when forces or accelerations are expressed in terms of $\mathbf{i}$ *and* $\mathbf{j}$, *it can be assumed that the units are N or m s^{-2}.*

10   A 6 kg particle accelerates at 3 m s^{-2} in the direction of the positive *y*-axis. Express the force on it in terms of $\mathbf{i}$ and $\mathbf{j}$.

11   A 40 N force acts on a 2 kg particle in the direction of the vector $3\mathbf{i} - 4\mathbf{j}$. Express the acceleration in terms of $\mathbf{i}$ and $\mathbf{j}$.

12   An 8 kg particle starts at the origin, at rest, and moves under the influence of a constant force to the point $(-1, 0)$ in 2 s. Express the force in terms of $\mathbf{i}$ and $\mathbf{j}$.

13   Two forces act on a 3 kg particle, causing it to accelerate at 4 m s^{-2} in the direction of the positive *x*-axis. If one of the forces is $7\mathbf{i} + 2\mathbf{j}$, what is the other?

14   Forces of $4\mathbf{i} - 9\mathbf{j}$, $2\mathbf{i} - \mathbf{j}$ and $-6\mathbf{i} + 4\mathbf{j}$ act on a 4 kg particle which is initially at rest at the point $(3, 5)$. Find its position and speed after 4 s.

15   Forces of 40 N and 26 N act on a 6 kg particle in the directions of the vectors $3\mathbf{i} + 4\mathbf{j}$ and $-12\mathbf{i} + 5\mathbf{j}$, respectively. Find the magnitude of the acceleration.

16   Three forces act on a 5 kg particle, causing it to accelerate in the direction of the vector $\mathbf{i} - \mathbf{j}$ at $6\sqrt{2}$ m s^{-2}. One of the forces has a magnitude of 25 N and is in the direction of the negative *y*-axis, while another has a magnitude of 50 N and is in the direction of the vector $24\mathbf{i} - 7\mathbf{j}$. Find the magnitude and inclination to the positive *x*-axis of the third force.

4

Forces Acting at a Point

Throughout this book we shall limit our attention to *coplanar* systems of forces, that is systems in which all the forces can be represented as acting in the plane of the paper we are using. In this chapter we begin by considering one of the simplest kinds of coplanar system, namely that in which all the lines of action pass through a single point, so that the forces can be described as *concurrent*.

The chapter starts with some analysis of systems of forces themselves, without reference to the bodies they act upon. The analysis normally consists in finding the *resultant* of a system, i.e. the simplest force or system to which it can be reduced. It will be seen that a concurrent system can always be reduced to a *single resultant force* (which may be zero). This however is not the case for all non-concurrent systems.

In the second part of the chapter we shall consider some systems of forces which act upon small objects. These objects will always be stationary and thus *in equilibrium*. The term 'equilibrium' can be applied both to objects and to systems of forces, and it may be formally defined as follows.

A system of forces is in equilibrium when it has a resultant of zero.
A body is in equilibrium when the resultant of the forces acting upon it is zero.

The parallelogram and triangle laws

It was pointed out in the last chapter that forces are localised vectors. They are vectors because they have magnitude and direction and because experiment shows that the resultant of two forces which act at a point can be obtained by the use of either the parallelogram or the triangle law for adding vectors. We thus have the following two laws or theorems for forces which correspond to these general vector laws.

The parallelogram of forces law

This states that if two forces acting at a point are represented in magnitude and direction by the directed line-segments $\overrightarrow{OA}$, $\overrightarrow{OB}$, their resultant is represented in magnitude and direction by $\overrightarrow{OC}$, where $OACB$ is a parallelogram (Fig. 4.1). Note the use of the double arrow-head to represent the resultant.

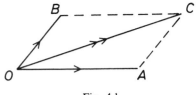

Fig. 4.1

The triangle of forces law

This law is customarily stated in a slightly different way to the general law for adding vectors, but it will easily be seen that the two forms are equivalent.

A system of forces in equilibrium can be represented by the sides of a triangle taken in order.

The law thus states that if in Fig. 4.2 **P**, **Q** and **S** constitute a system which is in equilibrium, the triangle ABC can be constructed such that $\overrightarrow{AB}$ represents **P** in magnitude and direction, and $\overrightarrow{BC}$, $\overrightarrow{CA}$ similarly represent **Q** and **S**.

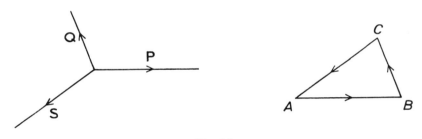

Fig. 4.2

The general triangle law for adding vectors would state that, for example, the resultant of **P** and **Q** (namely $-$**S**) could be obtained by constructing the same triangle ABC. Then $\overrightarrow{AC}$ would represent the resultant.

The line of action of a resultant

To specify a resultant force completely we need in general to state its magnitude, direction and line of action. In the special case of forces acting at a point the location of the line of action presents no problem: the resultant clearly must act through the point at which the original forces act. There is just one type of case in which confusion does sometimes arise with regard to the line of action of a resultant; this type is illustrated by the following example.

Suppose that forces act along two of the sides of a triangle ABC, the forces being *completely* represented – that is, in size, direction, and *line of action* – by $\overrightarrow{AB}$ and $\overrightarrow{BC}$, as shown in Fig. 4.3. By the triangle law the resultant is represented in size and direction by $\overrightarrow{AC}$, and it is tempting to suppose that it is completely represented by

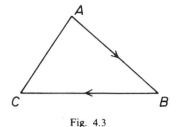

Fig. 4.3

$\overrightarrow{AC}$. In fact of course the resultant must act through the point of intersection of the lines of action of the original two forces, namely *B*. We thus must write

force $\overrightarrow{AB}$ + force $\overrightarrow{BC}$ = force represented by $\overrightarrow{AC}$, acting through *B*.

Methods for finding the resultant of any number of concurrent forces

We begin with the special case of two forces, which can be dealt with by the direct use of the parallelogram law in conjunction with the trigonometric sine and cosine formulae.

Example   Two forces of 5 N and 7 N act at a point. Both are directed away from the point, and the angle between them is 60°. Find the magnitude and direction of the resultant.

The parallelogram of forces is as shown in Fig. 4.4.

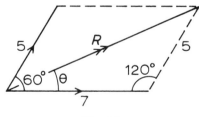

Fig. 4.4

By the cosine formula we have

$$R^2 = 7^2 + 5^2 - 2 \times 7 \times 5 \times \cos 120°$$

$$= 49 + 25 + 35 \quad (\text{using } \cos 120° = -\cos 60° = -\tfrac{1}{2})$$

$$= 109$$

∴ $R = 10.44$ N.

To find θ, and thus the direction of the resultant, we use the sine formula:

$$\frac{\sin \theta}{5} = \frac{\sin 120°}{10.44}$$

from which

$$\theta = 24.50°.$$

We proceed next to the general case of any number of concurrent forces. The method here is based on *resolution*, which was explained in the last chapter. Particularly important is the result obtained there for the value of a rectangular component, which, applied to the case of a force, becomes

---

component = original force × cosine of angle
between original force and component.

It is important to remember that a force must always be resolved into *two* rectangular components. (A common mistake is to forget one of them.) Also worth noting is that since $\cos(90° - \theta) = \sin\theta$, the two components of a force of magnitude F can always be expressed as $F\cos\theta$ and $F\sin\theta$. Figures 4.5–4.7 illustrate the resolution of a force.

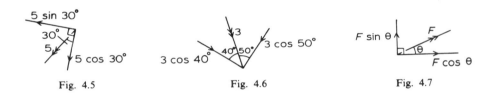

Fig. 4.5   Fig. 4.6   Fig. 4.7

Worked examples

Example 1   Horizontal forces of 15 N, 25 N and 20 N act away from a certain point on bearings of 45°, 180° and 300°. Find their resultant in magnitude and direction (Fig. 4.8).

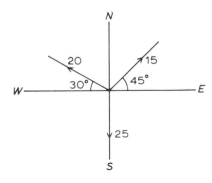

Fig. 4.8

We begin by choosing two directions at right angles and resolving all the forces into components in these directions. Here we obviously choose the *N–S* and *E–W* directions and obtain:

total component towards the south = $25 - 15 \cos 45° - 20 \cos 60°$

$$= 4.393;$$

total component towards the west = $20 \cos 30° - 15 \cos 45°$

$$= 6.714,$$

giving the situation shown in Fig. 4.9.

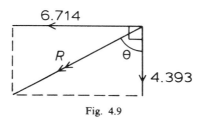

Fig. 4.9

By Pythagoras

$$R^2 = 6.714^2 + 4.393^2$$

from which

$$R = 8.024 \text{ N}.$$

Also

$$\tan \theta = \frac{6.714}{4.393}$$

from which

$$\theta = 56.80°.$$

(The actual bearing of the resultant is $56.80° + 180°$, i.e. $236.80°$.)

The resolution method can of course be applied to the case of two forces, and sometimes – as in the following example – is preferable to the parallelogram method.

Example 2   Find the resultant of two forces of 40 N, acting towards a certain point, which are inclined to one another at 130° (Fig. 4.10).

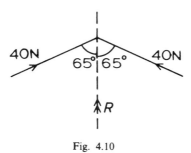

Fig. 4.10

By symmetry the resultant clearly acts along the axis of symmetry shown, and as the component in this direction of each force of 40 is 40 cos 65°, we have

$$R = 2 \times 40 \cos 65° = 80 \cos 65° = \mathbf{33.81 \ N}.$$

(The two components of $40 \sin 65°$, perpendicular to the axis of symmetry, cancel each other out.)

Related problems

In some problems we are given information about the resultant and some feature of the original system has to be found. The solution of such problems involves no new principles or techniques, as the following examples show.

Example 1   Forces of 3 N and 5 N, acting towards a certain point, have a resultant of 6 N. Find the angle between the two forces.

Here we draw a parallelogram of forces. Note that the parallelogram is best drawn, as previously, with all arrows directed *away* from their point of intersection (Figs. 4.11, 4.12).

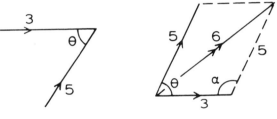

Fig. 4.11   Fig. 4.12

The method is to use the cosine formula to find α, and then use the fact that $\theta = 180° - \alpha$. We have

$$\cos \alpha = \frac{3^2 + 5^2 - 6^2}{2 \times 3 \times 5} = -\frac{1}{15}$$

from which

$$\alpha = 180° - 86.18° \quad \text{and} \quad \boldsymbol{\theta = 86.18°}.$$

Example 2   Horizontal forces of 15 and P act away from their point of application on bearings of 40° and 240°. Given that the resultant R is due west in direction, find P and R (Fig. 4.13).

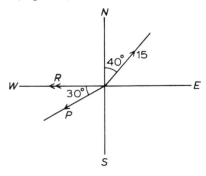

Fig. 4.13

In this problem the resolution technique is best. We note that since the direction of R is due west, the original system must have a net component of zero in the N–S direction. Hence

$$P \cos 60° = 15 \cos 40°, \qquad \therefore P = 30 \cos 40° = \mathbf{22.98}.$$

Now, knowing P, we can obtain R by resolving in the E–W direction:

$$R = 22.98 \cos 30° - 15 \cos 50° = \mathbf{10.26}.$$

Exercise 4a

1   Two forces, with magnitudes P and Q, are both directed away from a certain point, and the angle between them is θ. Find the magnitude of their resultant and its inclination to the force of P in the following cases:
(a) $P = 20, Q = 15, \theta = 70°$;   (b) $P = 8, Q = 8, \theta = 120°$;
(c) $P = 9, Q = 12, \theta = 90°$;   (d) $P = 27, Q = 46, \theta = 148°$.
2   Forces of 12 N and 16 N act away from a certain point and have a resultant of 24 N. Find the angle between the two forces.
3   Forces of 5 N and 7 N act towards a certain point and have a resultant of 5 N. Find the angle between the two forces.
4   Find the magnitudes and bearings of the resultants of the following systems of horizontal concurrent forces, all of which act away from their points of application:
(a) 15 N, 10 N, 12 N, on bearings of 90°, 180°, 300°;
(b) 4 N, 6 N, 4 N, 7 N, on bearings of 30°, 160°, 180°, 270°;
(c) 10 N, 15 N, 25 N, on bearings of 70°, 160°, 340°;
(d) 17 N, 8 N, 12 N, 23 N, on bearings of 65°, 180°, 260°, 324°.
5   Horizontal forces of 50 and P act away from a certain point on bearings of 90° and 210°. Given that the resultant R is due south in direction, find P and R.
6   Horizontal forces of P, 5 and Q act away from a certain point on bearings of 0°, 115° and 270°. Given that the system is in equilibrium, find P and Q.
7   Coplanar forces of 10 N, 16 N and 30 N act away from a certain point, the angle between each pair of forces being 120°. Find the magnitude of the resultant and its inclination to the force of 10 N.
8   Horizontal forces of 24, 37 and P act away from a certain point on bearings of 110°, 230° and 350°. Given that the resultant R is due west in direction, find P and R.
9   Horizontal forces of 4, P and 3 act away from a certain point on bearings of θ, 180° and 310°, the angle θ being between 0° and 90°. Given that the system is in equilibrium, find θ and P.
10   Horizontal forces of P, 10 and 25 act away from a certain point on bearings of 90°, 242° and θ, where θ is between 270° and 360°. Given that the resultant is a force of 5 which is due north in direction, find θ and P.
11   Coplanar forces of 18 N, 26 N, 22 N and 14 N act away from a certain point, the angles between the pairs of forces being respectively 120°, 150° and 50°. Find the magnitude of the resultant and its inclination to the force of 18 N.
12   A particle is pulled by two horizontal ropes which exert forces of 50 N and 30 N. If it starts to move at right angles to the force of 30 N, what is the angle between the ropes?

13   A 3 kg block, lying on the ground, is dragged along by a rope which applies a force of 40 N to it. If the normal reaction applied by the ground to the block is 10 N, what is the inclination of the rope to the ground?

14   A 6 kg particle, lying on smooth ground, is pulled by horizontal forces of P and Q acting on bearings of 54° and 180°. Given that the particle moves due east with an acceleration of 3 m s^{-2}, find P and Q.

15   Forces of 8 N and P, inclined at an acute angle of θ, act on a 2 kg particle. The result is that the particle accelerates at 5 m s^{-2} at an angle of 30° to the force of P. Find θ and P.

16   A certain force has a horizontal component of 8 N. When it is turned through 90°, in a vertical plane, its horizontal component becomes 5 N. Find the magnitude of the force.

Bodies in equilibrium under systems of concurrent forces

The bodies we are concerned with here – particles, small rings, etc. – are small enough for their own dimensions to be ignored. The forces acting on them must therefore be concurrent. Since in each case the resultant force on the body is zero, problems on this topic usually require us to find unknown forces on the body, or unknown directions.

The normal method of solution is to resolve all the forces into two convenient perpendicular directions, and use the fact that the algebraic sums of the components in both directions are zero. An alternative approach to *three-force problems only* is to draw a triangle of forces or use *Lami's theorem* (see p. 36) which is derived from the triangle theorem. In the worked examples which follow we shall concentrate principally on the resolution method since this is applicable in all cases.

Example 1   A 500 g particle stands on a smooth plane inclined at 60° to the horizontal. Find the least horizontal force which prevents it from sliding and the corresponding reaction between the particle and the plane.

Since the plane is smooth it can exert only a force which is normal to its surface. Note that the force diagram (Fig. 4.14) shows only forces *on the particle*; since the plane is immovable we are not concerned with the forces which the particle exerts on it. Resolving vertically and horizontally we have

$$R \cos 60° = 5 \quad (1)$$

$$R \cos 30° = P \quad (2)$$

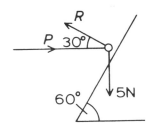

Fig. 4.14

Since $\cos 60° = \frac{1}{2}$, equation (1) gives $R = 10\,\text{N}$; and substituting this value in (2), we obtain $P = \textbf{8.660 N}$.

Example 2 A 2 kg particle lies on a smooth plane. Given that a force of 10 N, acting away from the plane at 20° to the line of greatest slope, is just sufficient to prevent it from sliding, find the angle of inclination of the plane and the reaction between it and the particle (Fig. 4.15).

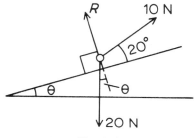

Fig. 4.15

Here we can obtain an equation in which the unknown R does not appear by resolving along the plane:

$$20 \sin \theta = 10 \cos 20°;$$

from which

$$\theta = \textbf{28.02}°.$$

Now to find R we resolve at right angles to the plane:

$$R + 10 \sin 20° = 20 \cos 28.02°$$

from which

$$R = \textbf{14.23 N}.$$

The next examples involve tensions in strings. There are two types of case to consider: that in which *separate* strings are attached to a particle, and that in which a ring is threaded onto a single string. In the first case the tensions are, in general, different; and in the second (assuming that there is no friction between ring and string) the tensions are the same throughout the string. Typical force diagrams are shown in Figs. 4.16, 4.17.

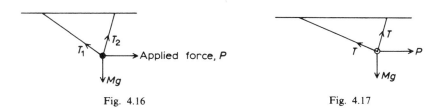

Fig. 4.16 Fig. 4.17

The two forces of T in the second diagram actually act on the small potion of string just below the ring, and this transmits the forces to the ring itself. The effect, however, is just as if the tensions acted directly on the ring, and this they are always in practice considered to do.

Example 3 A 5 kg particle is attached by two strings of lengths 60 cm and 80 cm of two points on a horizontal ceiling which are 1 m apart. Find the tensions in the strings (Fig. 4.18).

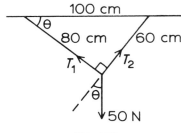

Fig. 4.18

Here it is important to notice that we have a 3,4,5 triangle. It follows that the strings meet at right angles. Resolving in the directions of the strings we obtain immediately

$$T_1 = 50 \sin \theta = 50 \times 3/5 = \textbf{30 N}.$$
$$T_2 = 50 \cos \theta = 50 \times 4/5 = \textbf{40 N}.$$

Example 4 A string of length 170 cm is attached at its ends to two points at the same level which are 130 cm apart. The string carries a smooth ring of mass 2 kg. A horizontal force is applied to the ring, causing the angle between the two sections of string to be 90°. Find this force and the tension in the string (Fig. 4.19).

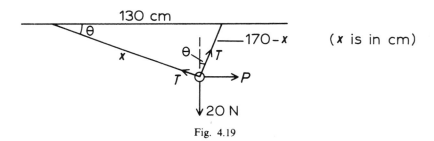

Fig. 4.19

Solution of the quadratic equation

$$x^2 + (170 - x)^2 = 130^2$$

gives $x = 50$ or 120. We thus have a 5,12,13 triangle and hence $\sin \theta = 5/13$ and $\cos \theta = 12/13$.

By resolving vertically we can obtain an equation which does not involve P:

$$T \sin \theta + T \cos \theta = 20,$$

that is,

$$T(5/13 + 12/13) = 20,$$

from which

$$T = \textbf{15.29 N}.$$

Now, knowing T, we resolve horizontally to find P:

$$P + T \sin \theta = T \cos \theta$$
$$\therefore \ P = 15.29(12/13 - 5/13)$$
$$\therefore \ P = 8.233 \text{ N}.$$

Example 5   A string $ABCDE$ in which $AB = DE$ and $BC = CD$ carries particles of mass 1 kg at B and D, and one of mass 2 kg at C. Its ends A and E are attached to two points on the same level, and the string hangs with BC and CD inclined at $60°$ to the vertical. Find the tensions in all the sections and the inclinations of AB and DE to the vertical (Fig. 4.20).

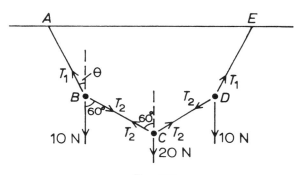

Fig. 4.20

By symmetry the tensions in AB and DE are equal, as are those in BC and CD.

We can find T_2 immediately by resolving vertically for the particle at C:

$$2 T_2 \cos 60° = 20$$
$$\therefore \ T_2 = 20 \text{ N}.$$

Now resolving horizontally and vertically for the particle at B gives two simultaneous equations in T_1 and θ:

Horizontally,

$$T_1 \sin \theta = T_2 \sin 60°;$$

that is,

$$T_1 \sin \theta = 10\sqrt{3} \qquad \text{(since } \sin 60° = \sqrt{3}/2\text{).} \qquad (1)$$

Vertically,

$$T_1 \cos \theta = T_2 \cos 60° + 10;$$

that is,

$$T_1 \cos \theta = 20. \qquad (2)$$

Dividing (1) by (2), we have $\tan \theta = \sqrt{3}/2$

$$\therefore \ \theta = 40.89°.$$

Substituting this value into equation (2), we obtain

$$T_1 = 26.46 \text{ N}.$$

Lami's theorem

This theorem applies to the special case of *three* concurrent forces in equilibrium, as shown in Fig. 4.21.

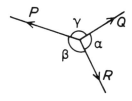

Fig. 4.21

By the triangle law, these forces can be represented in magnitude and direction by the three sides of a triangle taken in order, as shown in Fig. 4.22.

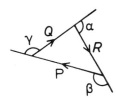

Fig. 4.22

Now applying the sine formula, and remembering that $\sin(180° - x) = \sin x$, we have

$$\frac{P}{\sin \alpha} = \frac{Q}{\sin \beta} = \frac{R}{\sin \gamma}$$

This is Lami's theorem. It should be noted that in the original diagram all the forces should be directed away from their point of application, or all towards it.

The type of problem in which Lami's theorem provides a significantly better method of solution than the resolution method is that in which there is no right angle between any pair of forces, and no other simplifying feature such as symmetry. The following example is of this kind.

Example 6 A string ABC has a body of mass 10 kg attached at B, and the ends A and C are attached to two points on a horizontal ceiling. If angles BAC and ACB are 50° and 60°, respectively, what are the tensions in AB and BC (Fig. 4.23)?

This problem can be done by resolving vertically and horizontally, but two somewhat awkward equations in T_1 and T_2 are then obtained. Lami's theorem on the other hand gives the simple equations

$$\frac{T_1}{\sin 150°} = \frac{T_2}{\sin 140°} = \frac{100}{\sin 70°}.$$

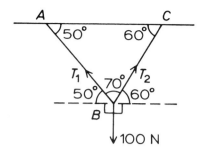

Fig. 4.23

We thus have

$$T_1 = \frac{100 \sin 150°}{\sin 70°} = \textbf{53.21 N}$$

and

$$T_2 = \frac{100 \sin 140°}{\sin 70°} = \textbf{68.40 N}.$$

Exercise 4b

1 A 4 kg particle lies on a smooth plane inclined at 30° to the horizontal. Find the force which just prevents sliding, and the corresponding normal reaction, if (a) the force acts straight up the plane, (b) the force is horizontal.

2 A 10 kg particle lies on a smooth plane. Find the inclination of the plane if (a) a force straight up the plane of 60 N, (b) a horizontal force of 120 N, is just sufficient to prevent sliding.

3 A particle on a smooth plane of inclination 40° is just held in equilibrium by a 20 N force which acts away from the plane and makes an angle of 30° with a line of greatest slope. Find the mass of the particle and the reaction between it and the plane.

4 A 2 kg particle lies on a smooth plane of inclination 25°, just being held in position by a force of P which acts away from the plane and makes an angle of 35° with a line of greatest slope. Find P and the normal reaction.

5 A 6.5 kg particle is attached by strings of length 50 cm and 120 cm to two points at the same level which are 130 cm apart. Find the tensions in the strings.

6 A 20 kg particle is attached by strings of length 150 cm and 2 m to two points at the same level which are 250 cm apart. Find the tensions.

7 A 500 g particle hangs on the end of a string. Find (a) the least force, (b) the horizontal force, which will cause the string to incline at 40° to the vertical. Find the corresponding tension in each case.

8 A string AC carries a smooth ring B of mass 800 g. The points A and C are attached to two points at the same level and a horizontal force of P is applied to the ring. If angles BAC and BCA are 60° and 30°, find P and the tension in the string.

9 A smooth ring of mass 140 g hangs in equilibrium at the centre of a string 50 cm long whose ends are attached to two points on a horizontal ceiling which are 48 cm apart. Find the tension in the string.

10   A string ABC carries a body of mass 2 kg at B, and A and C are attached to two points at the same level. If angles BAC and BCA are 30° and 70°, what are the tensions in AB and BC? (Use Lami's theorem.)

11   A string of length 21 cm is attached to two points on a ceiling which are 15 cm apart. A 250 g ring is threaded onto the string. If a horizontal force on the ring causes it to rest in equilibrium a distance 9 cm from one end of the string, find the value of the force and the tension in the string.

12   A 5 kg block, lying on the ground, is pulled by a rope which is inclined at 20° to the ground. The block just fails to move because the motion is opposed by a horizontal frictional force of 40 N. Find the tension in the rope and the normal reaction between the block and the ground.

13   A 4 kg block, lying on smooth ground, is pulled by two ropes in the same vertical plane which are inclined at 30° and 40° to the ground. If the block fails to move and the reaction of the ground is 10 N, what are the tensions in the ropes?

14   A 5 kg block is in equilibrium on the ground. A rope inclined at 45° to the ground pulls on it, and a horizontal frictional force opposes the attempted motion. The normal reaction of the ground is 20 N. Find the tension in the rope and the frictional force.

15   A string $ABCD$ in which $AB = CD$ has 300 g particles attached at B and C, and the ends A and D are attached to two points on the same level. Given that the tension in AB is 4 N, find the inclination of AB to the horizontal and the tension in BC.

16   A string $ABCD$ carries a particle of mass 2 kg at B and a smooth ring at C. The points A and D are attached to two points on the same level and owing to a horizontal force of 3 N on the ring the string hangs with BC horizontal. The tension in BCD is 18 N. Find the tension in AB, the inclination of CD to the vertical, and the mass of the ring.

17   Particles of weight W_1 and W_2 are attached to a string $ABCD$ at B and C, and the string is suspended with the ends A, D on the same level. The angles BAD and BCD are 60° and 90°, respectively, and the tensions in AB and CD are respectively 6 N and 10 N. Find the inclination of CD to the horizontal, the tension in BC, and the values of W_1 and W_2.

5

Work, Energy, Power

Work

Consider the effect of the force of gravity on any ordinary object such as a book.
Figures 5.1 and 5.2 show two possibilities.

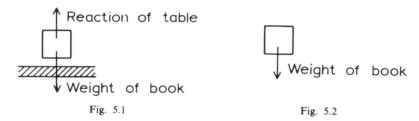

Fig. 5.1   Fig. 5.2

In Fig. 5.1 the book stands on a table, and this provides an upward force, which
balances the weight of the book and prevents it from falling. Consequently the
force of gravity has no observable *effect* on the book; it has no effect on the book's
position or velocity relative to the surrounding objects. In Fig. 5.2, on the other
hand, which shows the book freely falling, the force of gravity is having an effect;
it is changing the book's position and velocity in a clearly observable way.

This simple illustration of a familiar fact – that forces vary in the effects they
have on objects – leads us to the idea of *work*. A force, or an object applying a
force, is said to do work when there is some *movement*, in the direction of the
force, of the point at which the force is applied. Thus the force of gravity in Fig.
5.1, above, is not doing work; the force of gravity in Fig. 5.2 is doing work. The
formal definition is as follows.

> **The work done by a force = the magnitude of the force × the
> distance moved in the direction of the force.**

The definition means that work can be regarded as a quantitative measurement
of the effect a force has on the positions and velocities of objects in the world. Let
us consider the definition in more detail. First, it is clearly reasonable that *distance
moved* should be included in the definition, if this is to measure the effect of a
force. The *magnitude of the force* is needed in the definition because a large force
has a greater effect than a small one when both move an object the same distance.
Contrast for example the effect a train has in crashing into another train and

moving it 2 m with the effect a person produces if he moves a book 2 m. In the first case the effect – in terms of damage, speed of flying fragments, etc. – is enormous, while in the second it is relatively slight. The last part of the definition, '*in the direction of the force*', is clearly needed because any component of motion which is at right angles to a force is not directly caused by that force.

The relevance of *direction* in calculations of work done is illustrated by the following simple example.

Example A 6 kg body slides 4 m down a smooth plane inclined at 30° to the horizontal. Find the work done by the force of gravity on the body (Fig. 5.3).

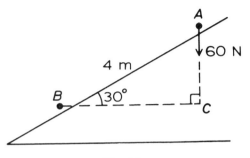

Fig. 5.3

Let the body start at *A* and finish at *B*. We can find the work done in two ways.
Method (a) Work done = force of 60 N × distance moved in the direction of this force

$$= 60\,\text{N} \times AC$$
$$= 60\,\text{N} \times AB \sin 30°$$
$$= 120\,\text{N m}.$$

Method (b) Resolving the force of 60 N into components along and perpendicular to the plane, we have the situation shown in Fig. 5.4. Now the component of 60 cos 30° is at right angles to the motion and thus does no work, while the component of 60 sin 30° is in the direction of the motion. Hence the work done is

$$(60 \sin 30° \times 4)\,\text{N m},$$

that is, **120 N m** as before.

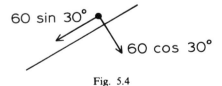

Fig. 5.4

The SI unit of work, as just indicated, is the N m. This is called the *joule* and denoted by J. The *kilojoule* is 1000 joules and denoted by kJ.

It should be noted that work is a *scalar* and not a vector quantity. The definition refers to the *magnitude* of the force and the *distance* moved (rather than the displacement), and these quantities are both scalars.

Energy

Whenever something actually does work, it is found that some of its ability to go on doing work is lost. The ability to do work is thus *used up* when work is done. Note that this does not happen when a force merely acts, and fails to cause movement. A compressed spring which is held in its compressed condition will go on applying a force indefinitely; once it is made to drive something, however, and thus do work, it loses its compression and thus uses up its ability to do work.

The ability to do work is called *energy*, and when work is actually done the amount of energy used is defined to be the amount of work done. Thus the units of energy are the same as those of work.

There are many kinds of energy, e.g. mechanical energy, heat energy, chemical energy, atomic energy; but here we shall be chiefly concerned with mechanical energy. This is of two kinds, namely *potential energy* (PE), which is energy due to position or configuration, and *kinetic energy* (KE), which is energy due to motion. Any moving body has the ability to do work because it can collide with another body and drive it along a certain distance before coming to rest. All moving bodies thus have KE. A compressed spring on the other hand has PE because the configuration of the coils enables the spring to apply a force to an object and move it. Also a raised body has PE because its position enables it to do work by applying a force to something else while falling.

Conservation of energy

Although something which is doing work is steadily losing its ability to do work and thus its energy, there is never an absolute loss of energy to the universe. For experience shows that whenever one thing loses energy, something else – normally that to which the force is applied when the work is done – gains exactly the same amount. This is well illustrated by the conversion of mechanical to heat energy, which occurs whenever a force moves an object against a frictional resistance. In such cases a very small amount of energy may be converted first to sound and then to heat, but the quantity of heat energy eventually produced is exactly equal to the work done in overcoming the resistance, and thus to the mechanical energy supplied. We have in fact the general principle:

Heat produced = frictional force × distance moved against frictional force.

The law governing energy conservation is expressed in a general way by the *principle of conservation of energy*, a simple statement of which is

Energy can neither be created nor destroyed.

Examples of energy conversion

1. A spring is initially held compressed, then released and allowed to accelerate a block. In this case the PE of the spring is converted mainly into the KE of the block. A little heat is also produced, owing to the action of air friction.

2.   A car is driven at constant speed. Since the speed is constant the force driving the car must be exactly balanced by frictional forces and the energy being supplied by the car's engine is thus entirely converted to heat.
3.   A ball is dropped to the ground, bounces a few times, and finally comes to rest. Initially the ball has gravitational PE. As it falls this is converted partly to KE and partly to heat owing to the action of air friction. Eventually all the energy is converted to heat.

Kinetic energy

We shall now derive an expression for the quantity of KE possessed by any moving body. This is simply the amount of work the body is capable of doing before coming to rest.

Consider a body of mass m which is moving with a speed of v. Let the body make contact with some other object and apply to it a constant force of F, in the direction of its motion. By Newton's third law it will then itself experience a retarding force of F, and thus have a deceleration of F/m. Let it travel a distance s before coming to rest.

We have: initial speed $= v$
final speed $= 0$
acceleration $= -F/m$
distance $= s$.

Now using

$$v^2 = u^2 + 2as \quad \text{(remembering that here } u = \textit{initial}\text{ speed and } v = \textit{final}\text{ speed)}$$

we obtain $0 = v^2 - 2Fs/m$

and hence $Fs = mv^2/2$.

This is the work that the body is capable of doing before coming to rest, and thus the energy it possesses when moving with a speed of v. Hence the result is

$$\boxed{\text{KE} = \frac{mv^2}{2}}$$

The expression $mv^2/2$ gives the KE of a body in joules when the mass is in kg and the speed is in m s^{-1}. Thus, for example, a 10 kg body moving with a speed of $6\,\text{m s}^{-1}$ possesses 180 J of KE.

Gravitational PE

As stated above, any *raised* body is capable of doing work by applying a force to something else while falling. Now consider a body of mass m which is raised to a height h. By the principle of conservation of energy, the PE the body possesses at this height is simply the work which has to be done *on* it to raise it. That is its own weight times its height, or mgh. We thus have the following result.

$$\boxed{\text{Gravitational PE} = mgh.}$$

Notes

1. It might be thought that a force slightly *above* a body's weight is required to lift it, and that a force equal to its weight merely prevents it from falling. This however is a mistake. An extra force may be needed to start the body off, but once it is moving it carries on at constant speed if there is no resultant force on it, i.e. if the upward force is *mg*. The work done by the extra force required to start the body is converted to *kinetic* energy, and should never be included in the expression for gravitational PE.

2. It will be noticed that we have spoken of raising a body, but have never specified the body's initial level. We can in fact take any convenient level as the datum level, relative to which gravitational PE is measured. Once this level is specified, the PE of bodies becomes positive, zero or negative according to whether they are above, at, or below the datum level. In problems there are two possibilities: the datum level can be clearly stated at the beginning; or – and this is often a more convenient procedure – we can refer only to *changes* in gravitational PE.

Energy in a stretched string or spring (or a compressed spring)

This is clearly another case of PE. To obtain the required expression we must first discuss *Hooke's law*, from which it is derived.

Consider an elastic string of natural length *l*, and suppose that a force is applied to it, along its length, which extends it a distance *e*. By Newton's third law the string must pull back with a force equal to the extending force; hence the extending force equals the tension in the string and can be denoted by *T* (Fig. 5.5).

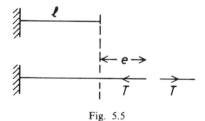

Fig. 5.5

Clearly if the extending force is large enough, it will damage the structure of the string (or spring). Provided that this is not so, experiment shows that *the extension is directly proportional to the extending force, and thus to the tension in the string.* This is known as Hooke's law, and it may be expressed mathematically as $T = ke$, where k is a constant of proportionality.

The value of the constant k depends on the material and length of the string, and the above equation shows that it equals the force required to extend the string by 1 unit of length. k thus measures, roughly speaking, the *strength* of the string. The idea of an elastic string's 'strength' (i.e. the difficulty that is experienced in stretching it) is important, but what is really required is a quantity which measures the strength of the *material* of a string, rather than one which varies with the length of particular samples. A quantity of this kind is obtained by re-expressing k as λ/l, where l is the string's natural length. The new constant λ is then a measurement of the strength of the material of the string and it is known as the material's *modulus of elasticity.*

In its most common form, then, Hooke's law is expressed as

$$T = \frac{\lambda e}{l}$$

The potential energy of a stretched string is equal to the work done in producing the extension. This can be found by integration – an integral being regarded as the *limit of a sum.* Consider a string of natural length l and modulus λ, and suppose we require the work done in producing an extension of e.

At an extension of x,

$$T = \frac{\lambda x}{l}, \qquad \text{by Hooke's law.}$$

Hence work done in stretching the string a further distance dx is

$$\frac{\lambda x \, dx}{l}.$$

Hence the total work done in producing an extension of e is

$$\int_0^e \frac{\lambda x \, dx}{l}$$

$$= \left[\frac{\lambda x^2}{2l} \right]_0^e$$

$$= \frac{\lambda e^2}{2l}.$$

The result is thus

$$\textbf{PE stored in a stretched string or spring} = \frac{\lambda e^2}{2l}$$

Worked examples

Example 1   A block which is initially at rest on a rough horizontal surface is pushed along a distance of 2 m by a force of 45 N. The frictional force, which directly opposes the motion, is 25 N. Find (a) the total work done, (b) the energy converted to heat, (c) the final kinetic energy of the block (Fig. 5.6).

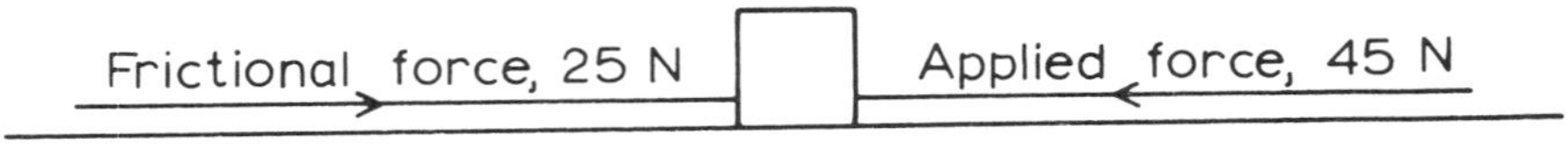

Frictional force, 25 N   Applied force, 45 N

Fig. 5.6

(a) The total work done is the work done by the *applied* force of 45 N, and it is therefore given by the product of this force and the distance moved. (A common mistake is to suppose that the total work done is obtained by multiplying the *resultant* force by the distance.) We have:

$$\text{total work done} = 45 \times 2 = \mathbf{90\ J}.$$

(b) The energy converted to heat is simply the work done against friction, i.e. the frictional force multiplied by the distance. Thus:

$$\text{energy converted to heat} = 25 \times 2 = \mathbf{50\ J}.$$

(c) The total energy supplied is the total work done, and all of this is converted either to heat or to the KE of the block. (No energy is lost, by the principle of conservation of energy.) Thus:

$$\text{final KE of block} = \text{energy supplied} - \text{energy converted to heat}$$
$$= 90\ J - 50\ J$$
$$= \mathbf{40\ J}.$$

Note that this result can also be obtained by multiplying the *resultant* force (20 N) by the distance. This is not surprising, for the resultant force can be regarded as the force which produces acceleration (we use it in the equation $F = ma$), and thus the work it does is converted to kinetic or 'motional' energy.

Example 2   A 40 kg body is lifted 3 m and thrown with a speed of 8 m s^{-1}. Owing to the action of air resistance, etc., it finally comes to rest on the ground. How much heat is produced?

The heat produced is equal to the mechanical energy supplied. That is,

$$\text{heat} = mgh + \tfrac{1}{2}mv^2$$
$$= (40 \times 10 \times 3) + (\tfrac{1}{2} \times 40 \times 64)$$
$$= \mathbf{2480\ J}.$$

Example 3   A child's slide is 8 m high and 15 m long. A 30 kg boy starts at the top, at rest, and slides down. If the average frictional force he experiences is 155 N, what is his speed at the bottom?

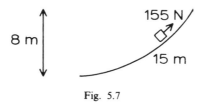

Fig. 5.7

Here it is best to start by writing down an *energy equation*, showing the relationship between the different forms of work and energy involved. In this case the equation is

loss of gravitational PE = gain of KE + work done against friction.

Using the standard formulae, this becomes

$$mgh = \tfrac{1}{2}mv^2 + Fs$$

that is,
$$30 \times 10 \times 8 = 15v^2 + (155 \times 15),$$
from which
$$v = \sqrt{5} = \textbf{2.236 m s}^{-1}.$$

Example 4   One end of an elastic string of natural length 25 cm is fixed to a point A on a ceiling. A 5 kg body, attached to the other end, is held at A and released. If it first comes to rest 50 cm below A, what is the modulus of elasticity of the string (Figs. 5.8, 5.9)?

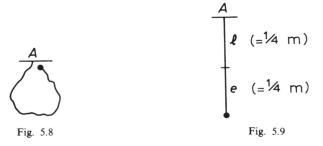

Fig. 5.8   Fig. 5.9

Since the body starts and finishes at rest, there is no net change in KE. The energy equation is thus

loss in gravitational PE = gain in energy stored in string.

Using the standard expressions proved earlier, this becomes

$$mgh = \frac{\lambda e^2}{2l}.$$

Now h is the total distance fallen by the body, i.e. $\frac{1}{2}$ m, and $e = l = \frac{1}{4}$ m. We thus have

$$5 \times 10 \times \tfrac{1}{2} = \frac{\lambda \times 1/16}{2 \times 1/4}$$
$$\therefore\ 25 = \lambda/8$$
$$\therefore\ \ \lambda = 200.$$

It follows from Hooke's law, namely $T = \lambda e/l$, that the unit of λ is the same as that of T.
     Hence the modulus of elasticity is **200 N**.

Example 5   A non-elastic string of length 80 cm is attached to a fixed point at one end and carries a weight at the other. The string hangs vertically and the weight is given an initial horizontal velocity of v. Find v if the string turns through $60°$ before coming to rest (Fig. 5.10).

Here the energy equation is simply

gain of gravitational PE = loss of KE;

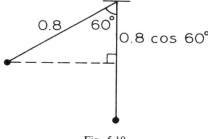

Fig. 5.10

that is,
$$mgh = mv^2/2$$
that is,
$$10(0.8 - 0.8 \cos 60°) = v^2/2 \qquad \text{(cancelling the } m\text{'s)}$$
from which
$$v^2 = 8$$
and
$$v = \mathbf{2.828\,m\,s^{-1}}.$$

Example 6 A pump raises 5 tonnes of water every minute from a depth of 25 m. The water is delivered through a pipe of cross-sectional area 50 cm². Find the work done by the pump per minute. (Ignore friction and take the density of water to be $1000\,\text{kg}\,\text{m}^{-3}$.)

First we find the work that is done simply in raising the water.

$$\text{Work done} = mgh$$
$$= 5000 \times 10 \times 25 \text{ J}$$
$$= 1250\,\text{kJ}.$$

In addition the water is given KE. To find this we need the speed of the water on emerging from the pipe, which can be obtained from the equation

$$\text{volume delivered per second} = \text{speed} \times \text{cross-sectional area}.$$

Now since volume = mass/density, we have

$$\text{volume delivered per minute} = 5000/1000 = 5\,\text{m}^3$$
$$\therefore \text{ volume delivered per second} = 5/60 = 1/12\,\text{m}^3.$$

Also the cross-sectional area $= 50\,\text{cm}^2 = 50/10^4\,\text{m}^2$.

Letting the speed of the water be $v\,\text{m}\,\text{s}^{-1}$, we thus have

$$\frac{1}{12} = v \times \frac{50}{10^4},$$

from which $v = 100/6$.

The KE supplied to the water per minute can now be found:

$$\text{KE supplied} = \tfrac{1}{2}mv^2$$
$$= \tfrac{1}{2} \times 5000 \times \frac{100^2}{6^2} \text{ J}$$
$$= 694.4\,\text{kJ}.$$

The total energy supplied per minute, and thus the total work done in this time, is therefore $1250 + 694.4 = $ **1944 kJ** (to 4 significant figures).

Exercise 5a

1   A force of 20 N moves a body 4 m against a frictional resistance of 5 N. Find (a) the total work done, (b) the gain of KE of the body.

2   A force of 70 N acts vertically upwards on a 5 kg body which is initially at rest. Find (a) the work done by this force in 2 s, (b) the gravitational PE gained in this time.

3   A 20 kg body slides 8 m down a smooth slope of inclination $40°$. Find the work done on it by gravity.

4   Find the KE of (a) a 25 kg body moving at $2 \, \mathrm{m \, s^{-1}}$, (b) a 500 g body moving at $72 \, \mathrm{km \, h^{-1}}$.

5   A 200 g ball is thrown with a speed of $30 \, \mathrm{m \, s^{-1}}$ at a height of 25 m. It finally comes to rest on the ground, all its mechanical energy having been converted to heat. Find the amount of heat produced.

6   A 2 tonne car starts at rest and freewheels down a hill of height 50 m, arriving at the bottom with a speed of $20 \, \mathrm{m \, s^{-1}}$. If all the original mechanical energy is converted either to KE or to heat, how much heat is produced?

7   A 6 kg block slides down a hill of height 20 m, travelling 60 m altogether. If it starts at rest and finishes with a speed of $5 \, \mathrm{m \, s^{-1}}$, what is the average frictional resistance of the hill?

8   A 5 kg block is initially at rest on a smooth slope of inclination $30°$. It is pushed straight up the slope by a force of 40 N, acting for 4 s. Find (a) the gain of gravitational PE, (b) the gain of KE.

9   A 4 kg body falls at a constant speed of $50 \, \mathrm{cm \, s^{-1}}$ through a viscous liquid. Find the heat produced in 6 s.

10   A non-elastic string of length 50 cm is attached at one end to a fixed point, and it carries a weight at the other. The weight is released when the string is inclined at $50°$ to the downward vertical. Find its speed at its lowest point.

11   An elastic string of natural length 75 cm and modulus 150 N is attached at one end to a point on a smooth horizontal table, and at the other to a 2 kg body. The body is pulled to a point on the table which is 125 cm from the fixed point, and released. Find its speed at the moment when the string becomes slack.

12   Repeat question 11, but now let the table provide a frictional force, directly opposing the motion, of 10 N.

13   A particle hanging at the end of a vertical non-elastic string of length 80 cm is given a horizontal velocity of $2 \, \mathrm{m \, s^{-1}}$. Find the inclination of the string to the vertical when it comes to instantaneous rest.

14   A particle hanging at the end of a vertical non-elastic string of length l is given a horizontal velocity of $\sqrt{lg}$. Find the angle the string turns through before coming to instantaneous rest.

15   A non-elastic string of length 60 cm is attached at one end to a fixed point A, and at the other to a 100 g particle. The particle is held level with A, with the string fully extended, and given a downward velocity of $2 \, \mathrm{m \, s^{-1}}$. It swings through a viscous liquid and comes to rest when the string has turned through $150°$. Find the average frictional resistance of the liquid.

16 One end of an elastic string of natural length 50 cm and modulus 150 N is attached to a fixed point *P*. A body attached to the other end is held at *P* and dropped. Find the mass of the body if its lowest point in the subsequent motion is 75 cm below *P*.

17 An elastic string of natural length 50 cm and modulus 60 N is attached at its upper end to a point *A* on a smooth plane inclined at arc tan $\frac{4}{3}$ to the horizontal. The other end is attached to a 500 g body. The body is held on the line of greatest slope through *A*, at a distance of 1 m from *A*, and released. Find its speed when it reaches *A*.

18 A nail of mass 5 g is driven horizontally into a piece of wood by a blow from a hammer. The hammer gives it an initial velocity of $2 \, \text{m s}^{-1}$ and it penetrates a distance of 0.25 cm. Find the average resistance of the wood.

19 A 5 tonne body is dropped from a height of 2 m and penetrates 10 cm into soft ground. Find the average resistance of the ground.

20 A 20 kg body slides down a rough slope, falling through a vertical distant of 5 m. It starts at rest and finishes at $4 \, \text{m s}^{-1}$. If it experiences an average frictional resistance of 100 N, what is the distance it actually travels?

21 Find the work done by a pump in each second, if in this time it drives 6 kg of water, initially at rest, through an orifice of area $40 \, \text{cm}^2$.

22 Find the work done per second by a pump which drives water which is initially at rest through an orifice of area $15 \, \text{cm}^2$ at a speed of $10 \, \text{m s}^{-1}$.

23 Every 25 s a pump raises $2 \, \text{m}^3$ of water from a depth of 20 m and delivers it through a pipe. Given that the pump supplies 25 kJ of energy every second, find the final speed of the water and the cross-sectional area of the pipe.

Power

The definition of power is

> **power = rate of doing work.**

The SI unit of power is the joule per second, known as the *watt* (W). Also much used is the *kilowatt* (kW) which is 1000 watts.

We customarily refer to the power of cars, engines, etc., which are devices for converting chemical, electrical or nuclear energy (for example) into mechanical work. A *powerful* engine is one that performs this conversion rapidly, and thus provides a large amount of mechanical energy in a given time.

An important equation

When a car, train or other vehicle is driven along, work has to be done against retarding forces such as friction and gravity. The force which does this work, usually supplied by an engine inside the vehicle, is known as a *tractive* force. When we speak of the *power* of, say, a car, we are referring to the work that can be done per second by the car's tractive force. It is this work which is ultimately provided by the chemical energy of the car's fuel.

Now since

power = work done per unit time

= tractive force × distance moved per unit time,

we have the equation

> **power = tractive force × speed**

In most cases the tractive force is of course a frictional force, namely the force between the road or rails and the wheels driven by the engine.

When a car or train is going downhill, the component down the hill of its own weight assists the motion. Nevertheless *the component of the weight is not considered to be part of the tractive force*. This is because we want the equation

power = tractive force × speed

to give the power of the engine or other agency which permanently drives the vehicle, and this will only be so if we interpret 'tractive force' as the force which such a permanent agency provides.

Figure 5.11 shows the forces in the direction of the motion which act in the typical case of a car or train being driven up a hill. It is not necessary in problems on power, which are dynamics problems, to show other forces such as the weight and normal reaction which would certainly be shown in statics problems.

In addition to the forces shown, we often need to refer to the following forces, which are derived from them:

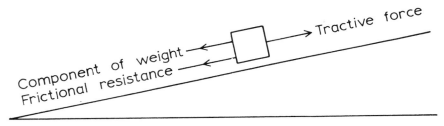

Fig. 5.11

Total retarding force. This is simply the sum of all the forces opposing the motion. In Fig. 5.11, it is the sum of the frictional resistance and the component of the weight. ('Retarding force' should be carefully distinguished from 'resistance'; the latter is just one among many kinds of retarding force.)

Total driving force. This is the sum of all the forces assisting the motion. Above, it is simply equal to the tractive force, but if the motion is downhill the component of the weight must be added to the tractive force.

Resultant force. This is clearly equal to the total driving force minus the total retarding force. It should be noted that the resultant force must be used in the equation $F = ma$.

Problems on power typically concern the motion of vehicles driven by engines, and they can normally be solved by the use of some or all of the following equations.

> **Power = tractive force × speed.**
>
> **Resultant force = mass × acceleration.**
>
> **Resultant force = total driving force − total retarding force.**

Worked examples on power

Example 1 A 400 kg vehicle fitted with a 3 kW engine is travelling up a gradient of 1 in 8 at maximum speed. Given that the frictional resistance is 100 N, find this speed.

A gradient of 1 in 8 means that $\sin \theta = 1/8$, where θ is the inclination of the plane to the horizontal. Since the component down the plane of a weight of W is $W \sin \theta$, we obtain this component by simply dividing the weight by 8. The relevant forces on the vehicle are thus as shown in Fig. 5.12.

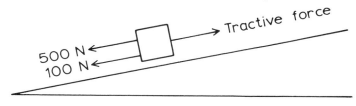

Fig. 5.12

When the vehicle reaches its maximum speed, it is no longer accelerating and thus the speed is constant. Hence there is no resultant force on the vehicle and we have

$$\text{tractive force} = 500 + 100 = 600 \text{ N}.$$

Now using the equation

$$\text{power} = \text{tractive force} \times \text{speed},$$

we have

$$3000 = 600 \times \text{speed},$$

from which the maximum speed is **50 m s⁻¹**.

Example 2 A car of mass 2000 kg has a maximum speed on a level road of 50 m s⁻¹, at which speed the total resistance is 800 N. Find the acceleration when the speed is 20 m s⁻¹, if the resistance is then 150 N.

Since the maximum speed is a constant speed, the tractive force at this speed = the total resistance = 800 N. Hence

$$\text{power of car} = \text{tractive force} \times \text{speed}$$
$$= 800 \times 50 = 40\,000 \text{ W}.$$

To find the tractive force when the speed is 20 m s⁻¹, we use the same equation again:

$$40\,000 = \text{tractive force} \times 20$$
$$\therefore \text{ tractive force} = 2000 \text{ N}.$$

Now to obtain the acceleration at this speed we need the resultant force, which is given by

$$\text{resultant force} = \text{tractive force} - \text{resistance}$$
$$= 2000 - 150 = 1850\,\text{N}.$$

Finally we use the equation $F = ma$, which gives

$$a = 1850/2000 = 0.925,$$

and the acceleration when the speed is $20\,\text{m}\,\text{s}^{-1}$ is thus **$0.925\,\text{m}\,\text{s}^{-2}$**.

Example 3   A car working with a constant power P against a constant resistance has a maximum speed of v when travelling up a slope of 1 in 12, and a maximum speed of $3v$ on the level. Obtain expressions in terms of P, v and g for the resistance and the mass of the car, and show that the acceleration when the car is travelling down the slope at a speed of v is $g/6$.

Using

$$\text{power} = \text{tractive force} \times \text{speed},$$

we have

$$\text{tractive force up the slope} = P/v,$$

and

$$\text{tractive force on the level} = P/3v.$$

Hence if the resistance is R and the mass is M, the forces on the car are as shown in Figs. 5.13 and 5.14.

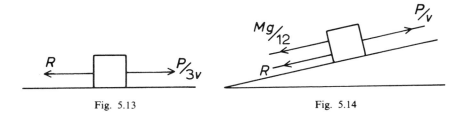

Fig. 5.13   Fig. 5.14

Since the speeds in both cases are constant, we have

$$R = P/3v$$

and

$$P/v = R + Mg/12.$$

Eliminating R:

$$P/v = P/3v + Mg/12$$
$$\therefore \ Mg/12 = 2P/3v$$
$$\therefore \ \ M = 8P/gv.$$

When the car is travelling down the slope at speed v, the tractive force is P/v. Hence

$$\text{total driving force} = P/v + Mg/12.$$

But

$$Mg/12 = 2P/3v \qquad \text{(see above)}$$
$$\therefore \ \text{total driving force} = 5P/3v.$$

Also

$$\text{resistance} = P/3v$$
$$\therefore \text{ resultant force} = 5P/3v - P/3v$$
$$= 4P/3v$$
$$\therefore \text{ acceleration} = \text{resultant force/mass}$$
$$= 4P/3v \times gv/8P$$
$$= g/6.$$

Example 4 A 200 tonne train can travel up a slope of 1 in 100 at a maximum speed of $10 \, \text{m s}^{-1}$. It has the same speed when freewheeling down the slope. Find the power of the train and the resistance at this speed. If the resistance is proportional to the square of the speed, find the maximum speed on the level.

Since 200 tonnes $= 200\,000 \, \text{kg}$, the weight of the train is $2\,000\,000 \, \text{N}$ and the component of the weight is $20\,000 \, \text{N}$. The relevant forces on the train are thus as shown in Figs. 5.15 and 5.16.

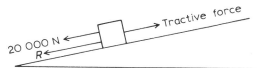

Fig. 5.15 (train moving up)

Fig. 5.16 (train moving down)

Since the speeds are constant the resultant force is zero in both cases. From the second diagram we therefore obtain immediately

$$R = 20\,000 \, \text{N}.$$

Applying this result to the first diagram, we have

$$\text{tractive force} = 40\,000 \, \text{N}.$$

Since power $=$ tractive force $\times$ speed, it follows that

$$\text{power} = 40\,000 \times 10$$
$$= 400\,000 \, \text{W} \quad \text{or} \quad \textbf{400\,kW}.$$

Now since the resistance is proportional to the square of the speed, we can write

$$R = kv^2.$$

The constant k can be found since we know that $R = 20\,000$ when $v = 10$:

$$k = 20\,000/100 = 200.$$

Now let the maximum speed on the level be $u \, \text{m s}^{-1}$. We can obtain an equation in u by expressing both the tractive force and the resistance in terms of u and using the fact that these are equal when the speed is constant.

$$\text{Tractive force} = \text{power/speed} = 400\,000/u;$$

and

$$\text{resistance} = 200\,u^2.$$

Hence
$$200\,u^2 = 400\,000/u$$
from which
$$u^3 = 2000.$$

It follows that the maximum speed on the level is **12.6 m s^{-1}**.

Exercise 5b

1 A car is working at 20 kW to maintain a steady speed of 20 m s^{-1} on a level road. Find the frictional resistance.

2 A car of mass 1 tonne is driven up a slope of 1 in 20 at a steady speed of 36 km h^{-1}. Find the power at which the car is working if the frictional resistance is negligible.

3 A 1.5 tonne car has a maximum power of 50 kW. If it moves against a constant resistance of 1000 N find its maximum speed (a) on the level, (b) up a gradient of 1 in 25.

4 A car of mass 2000 kg can freewheel down a slope of 1 in 10 at 15 m s^{-1}. Find the power needed to maintain the same speed on the level, if the resistance remains the same.

5 A train of mass 250 tonnes can develop a power of 4000 kW. Find its maximum speed up a gradient of 1 in 20 if the frictional resistance is then $\frac{1}{40}$ of the train's weight.

6 A train of mass 400 tonnes can develop 6000 kW. It has the same maximum speed on the level as when freewheeling down a slope of 1 in 20. Assuming that the frictional resistance is the same in both cases, find this speed.

7 A train of mass 100 tonnes is accelerating at 0.5 m s^{-2}. Find the resultant force on the train. If this acceleration occurs when the train is travelling up a slope of 1 in 25, against a negligible resistance, what tractive force is being exerted?

8 A car of mass 1.4 tonnes is accelerating at 2 m s^{-2}. This occurs when the speed is 10 m s^{-1} and the total retarding force is 1000 N. Find the resultant force, the tractive force and the power.

9 A train of mass 150 tonnes and power 2000 kW has a maximum speed of 20 m s^{-1} up a slope of 1 in 75. What acceleration does it have at the same speed on the level, if the power and resistance remain the same?

10 Find the power of a car of mass 1.5 tonnes if its acceleration is 0.6 m s^{-2} up a slope of 1 in 20 when the speed is 10 m s^{-1} and the resistance is $\frac{1}{10}$ of its own weight. Find its acceleration down the same slope, if power, speed and resistance are unaltered.

11 An engine of mass 50 tonnes which is developing 500 kW pulls a train of mass 200 tonnes against retarding forces amounting to 100 N per tonne. The train travels at uniform speed. Find the speed and the tension in the coupling.

12 A 1.5 tonne car works at 40 kW in towing a 500 kg caravan at 72 km h^{-1} up a gradient of 1 in 40. The frictional resistances are $\frac{1}{20}$ of the weight of each vehicle. Find the acceleration and the tension in the coupling.

13 A car of mass 1.6 tonnes and power 50 kW can accelerate on the level at 1.5 m s^{-2} when its speed is 20 m s^{-1}. Find its maximum speed on the level if the resistance is proportional to the square of the speed.

14 A car is driven on the level with a constant power against a constant resistance. When the speed is increased from v to $2v$, the acceleration drops to a third of its original value. Find the maximum speed in terms of v.

15 A car working at a constant power against a constant resistance has a maximum speed of $3v$ up a slope of 1 in 10, and a maximum speed of $2v$ up a slope of 1 in 5. Find its maximum speed on the level, in terms of v.

16 A 350 tonne train has a maximum speed of 108 km h^{-1} when travelling down a slope of 1 in 50. Its maximum speed is 54 km h^{-1} when it travels up the same slope. Given that the resistance is proportional to the square of the speed, find the power of the train and its maximum speed on the level.

17 A car of weight W and power P has a maximum speed of v when travelling up a slope of 1 in N, and a maximum speed of $2v$ when travelling down the slope. In the first case the resistance is R, and the resistance is proportional to the square of the speed. Express (a) P in terms of R and v; (b) R in terms of W and N.

18 A 1 tonne car pulls a 2 tonne caravan up a gradient of 1 in 20 at 20 m s^{-1}. The frictional resistances are 1/40 of the weight of each vehicle and the tension in the coupling is 1700 N. Find the acceleration and the power being supplied.

19 A certain car has a maximum speed of v when travelling up a slope of 1 in N, and a maximum speed of $3v$ when travelling down the slope. Prove that if the resistance is proportional to the square of the speed, the car's maximum acceleration when travelling at a speed of $3v/2$ on the level is $11g/26N$.

20 A car can pull a trailer of twice its mass up a certain slope at a maximum speed of v. The resistance per unit mass is proportional to the square of the speed. Given that when the trailer becomes detached the maximum speed of the car rises to $2v$, prove that the constant speed at which the car will freewheel down the slope is $v\sqrt{5}$.

6

Moments and Couples. Parallel Forces

The moment of a force

Suppose a body is mounted so as to be free to turn about some axis. A force will have a tendency to turn the body if (a) its line of action does not pass through the axis, and (b) it is perpendicular to the axis or has a component perpendicular to it. The force's turning effect depends both upon the magnitude of the force and its perpendicular distance from the axis, and we therefore obtain a measurement of its turning effect, or *moment*, by taking the product of these two quantities.

In practice, as we have seen, forces are usually represented as acting in the plane of the paper being used, and the axes about which we require their turning effects are usually then perpendicular to the paper. Such axes are located by points on the paper, and in these circumstances it is customary to refer to *moments about points* rather than moments about axes. The definition is as follows.

> **The moment of a force about a point is the magnitude of the force multiplied by the perpendicular distance of the point from the line of action of the force.**

It follows that a force has no moment about a point on its line of action.

Directions of moments

When the forces are in the plane of the paper the directions of their moments are conveniently specified as either *clockwise* or *anticlockwise*. Sometimes the convention 'anticlockwise positive, clockwise negative' is used; but if so, this should be stated. **The SI unit of moment is the N m.**

Two theorems on moments

We state these without proof.

1. *The algebraic sum of the moments of a system of (coplanar) forces about any point is equal to the moment of their resultant about that point.*

2. This is called the *principle of moments*, and it follows from (1):
If a system of (coplanar) forces is in equilibrium, the algebraic sum of the moments about every point in their plane is zero.

(The algebraic sum of the moments of a system of forces about a point is usually called the *resultant moment* about that point.)

A simple example is now given to illustrate the calculation of moments about various points.

Example The following diagram shows a rectangle of length 8 m and breadth 6 m. Find the resultant moment about *A*, *C* and *D*.

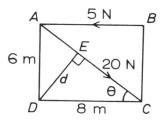

Fig. 6.1

The lines of action of both forces pass through *A*; hence

moment about *A* = zero.

The line of action of the 20 N force passes through *C*, so the moment of this force about *C* is zero. The perpendicular distance of the 5 N force from *C* is 6 m, and this force would tend to turn a body anticlockwise if it was mounted on an axis through *C*. Hence

moment about *C* = 5×6 = 30 N m anticlockwise.

There are two methods for calculating the moment of the 20 N force about *D*. We can begin either by calculating the perpendicular distance *d*, or by resolving the 20 N force into two perpendicular components at some convenient point on its line of action. One of these components should, if possible, pass through *D*.

Method (a) From triangle *CED*, $d = 8 \sin \theta$; and since *ACD* is a 3,4,5 triangle, $\sin \theta = \frac{3}{5}$. Hence $d = \frac{24}{5}$ and the moment of the 20 N force about *D* is $20 \times \frac{24}{5}$ = **96 N m clockwise.**

Method (b) We resolve the 20 N force at *A* (*C* would be equally suitable), obtaining a component acting through *D* together with a component along *AB* of $20 \cos \theta$. Hence moment of 20 N force about $D = 20 \cos \theta \times 6 = 20 \times \frac{4}{5} \times 6$ = **96 N m clockwise** once more.

Since the moment of the 5 N force about *D* is $5 \times 6 = 30$ N m anticlockwise, we have

resultant moment about *D* = 66 N m clockwise.

Couples

Definition:

> **A couple is a system consisting of two equal and opposite forces with a different line of action.**

Important property   A couple has the same moment about every point.

Proof   The moment of the couple shown in Fig. 6.2 about any point A is

$$F(d + x) - Fx = Fd \text{ anticlockwise.}$$

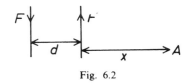

Fig. 6.2

This is an important result which may be stated in words as follows.

> **Moment of a couple = one of its forces × distance apart of the forces.**

A couple in a given plane is completely specified by its moment; there is no need, in stating its value, to give the actual size, direction or distance apart of its two forces. The three couples shown in Fig. 6.3, for example, are all equivalent, each being describable simply as a clockwise couple of moment 24 N m.

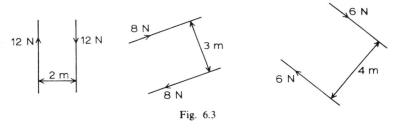

Fig. 6.3

It should be noted that a couple has no resultant force in any direction. This suggests that a couple can have no tendency to move a body as a whole, and in fact it can be shown that however a couple is applied to a body it has no effect on the centre of gravity, merely tending to turn the body about this point.

A couple is a *basic unit* in mechanics; it cannot be reduced to anything simpler. Thus if a system of forces is equivalent to a couple, that couple is its *resultant*.

Combination of Couples

Provided that they are coplanar, it is easily shown that the moments of couples can simply be added or subtracted to obtain resultant couples.

Two theorems on couples

1. *Any force, together with a coplanar couple, is equivalent to an equal parallel force.*

Proof   Let the force and couple have values P and L, and let the couple be expressed as two forces of magnitude P, parallel to the given force, a distance L/P

apart. We can then position the couple in such a way that one of its forces nullifies the given force (Fig. 6.4). Clearly the effect is to produce a single force of magnitude P, parallel to the force originally given.

The second theorem is the converse of the first.

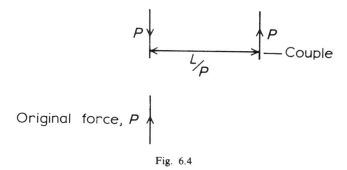

Fig. 6.4

2. *Any force is equivalent to an equal parallel force together with a couple.*

Proof At some point which is not on the line of action of the original force, add two opposite forces, both equal in magnitude and parallel to the original force, as shown in Fig. 6.5. (This amounts to adding zero.) We now have a couple – consisting of the upper two forces in Fig. 6.5 – together with an equal parallel force.

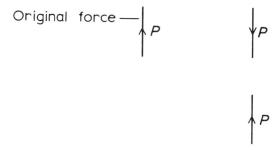

Fig. 6.5

Resultants of systems of parallel forces

To specify any resultant force completely we need to state its *magnitude, direction* and *line of action*. In the case of parallel forces the first two of these can be obtained immediately: the direction is that of the original forces and the magnitude is their algebraic sum. To find the line of action of the resultant we use the first of the two theorems on moments stated above, which may be abbreviated to the following important principle:

> **Moment of resultant = moment of original system (about any point).**

The special case in which the algebraic sum of the forces is zero is dealt with in the third of the worked examples which now follow.

Example 1 Find the resultant of the system shown in Fig. 6.6.

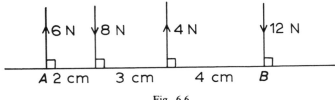

Fig. 6.6

The resultant is a force of 10 N in the downward direction. Let it cut AB a distance x to the right of A. Then using the principle

moment of resultant = moment of original system,

we have, taking moments about A:

$$10x = (8 \times 2) - (4 \times 5) + (12 \times 9)$$
$$= 104$$
$$\therefore \quad x = 10.4 \text{ cm.}$$

Example 2 An anticlockwise couple of 20 N cm is added to the system of Example 1. Find the new resultant.

The resultant is still a downward force of 10 N, since a force together with a couple is equivalent to an equal parallel force. The effect of the couple is simply to provide an additional anticlockwise moment of 20 N cm about every point; hence the above equation in x becomes

$$10x = 104 - 20$$
$$\therefore \quad x = 8.4 \text{ cm.}$$

Example 3 Show that the resultant of the system shown in Fig. 6.7 is a couple, and find its moment.

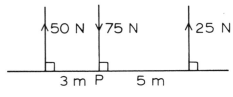

Fig. 6.7

Since the algebraic sum of the forces is zero the system does not reduce to a resultant force. It must therefore either reduce to a couple (the other basic unit) or be in equilibrium. Since in each case the system has the same moment about every point we need to calculate the moment about one point only: if the result is not zero the system cannot be in equilibrium and must be equivalent to a couple of the moment found.

Anticlockwise moment about $P = (25 \times 5) - (50 \times 3) = -25$.

The resultant is thus a **clockwise couple of moment 25 N m**.

Example 4 Forces of $-6\mathbf{i}$, $2\mathbf{i}$ and $F\mathbf{i}$ act at the points $(0, -2)$, $(0, 4)$ and $(0, a)$, as shown in Fig. 6.8. Given that the resultant is a clockwise couple of moment 30, find F and a. (The vector terminology used here is explained in Chapter 3.)

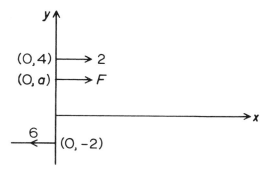

Fig. 6.8

Since the system is equivalent to a couple, the sum of the components in any direction is zero. Hence

$$F = 4.$$

To find a it is a best to take moments about the *origin*, as all the distances in the question are measured from this point.

$$(6 \times 2) + (2 \times 4) + Fa = 30$$

from which

$$a = 2.5.$$

Exercise 6a

Note: Parallel forces acting in the same direction are called *like* forces; parallel forces acting in opposite directions are called *unlike* forces.

1 Forces of 7 N, 5 N, 4 N act along the sides AB, CB, DA of a square $ABCD$ of side 10 cm (the directions being indicated by the order of the letters). Find the resultant moment about each of the points A, C and the centre of the square.
2 A force of 50 N acts along the diagonal AC of a rectangle $ABCD$ in which $AB = 3$ cm and $BC = 4$ cm. Find the moment of the force about D.
3 Forces of $2\mathbf{j}$, $-3\mathbf{i}$, $4\mathbf{i} - \mathbf{j}$ act at the points $(3,0)$, $(0,5)$, $(2,3)$. Find the resultant moment about the origin.
4 $ABCD$ is a rectangle in which $AB = 12$ cm and $AD = 5$ cm. Forces of 30 N, 15 N, 20 N, 26 N act along BA, DC, DA, BD. Find the resultant moment about each of the points B, A, C.

5   Forces of $-8\mathbf{i} + 5\mathbf{j}$ and $6\mathbf{i} - 2\mathbf{j}$ act at the origin and the point $(4,1)$. Find the resultant moment about the point $(3, -2)$.

6   Forces of $-4\mathbf{i} + P\mathbf{j}$ and $Q\mathbf{i} + 7\mathbf{j}$ act at the points $(3,0)$, $(-5,1)$. Find P and Q if the resultant moments about $(-5,1)$ and the origin are respectively 20 anticlockwise and 30 clockwise.

7   Forces of $6\mathbf{i} + \mathbf{j}$ and $-2\mathbf{i} + F\mathbf{j}$ act at the points $(a,3)$, $(-4,0)$. Find F and a if the clockwise moments about the origin and the point $(1,3)$ are respectively 31 and 25.

8   In each of the following cases find the magnitude of the resultant force and its perpendicular distance from A.

(a) AB is a line of length 20 cm. Like forces of 4 N and 6 N, perpendicular to AB, act at A and B, respectively.

(b) AB is a line of length 4 cm. Unlike forces of 3 N and 5 N, perpendicular to AB, act at A and B, respectively.

(c) ABC is a line in which $AB = 2$ cm and $BC = 3$ cm. Like forces of 10 N and 12 N at A and C, and an opposite force of 14 N at B, act at right angles to the line.

(d) $ABCD$ is a line in which $AB = BC = CD = 10$ cm. Forces of 8 N, -3 N, -6 N, 4 N act at right angles to the line at A, B, C, D, respectively.

(e) $ABCD$ is a line in which $AB = BC = 2a$ and $CD = 3a$. Forces of $4F$, $-5F$, $2F$, $-7F$ act at right angles to the line at A, B, C, D, respectively.

9   Forces of $5\mathbf{i}$, $10\mathbf{i}$, $-8\mathbf{i}$, $-7\mathbf{i}$ act at the points $(0,0)$, $(0,1)$, $(0,2)$, $(0,3)$. Show that the resultant is a couple and find its moment.

10   Forces of $2\mathbf{j}$, $5\mathbf{j}$, $-7\mathbf{j}$ act at the points $(-3,0)$, $(2,0)$, $(3,0)$. Find the resultant.

11   Forces of $12\mathbf{j}$, $-18\mathbf{j}$, $-9\mathbf{j}$, $15\mathbf{j}$ act at the points $(-1,0)$, $(3,0)$, $(6,0)$, $(8,0)$. Prove that the system is in equilibrium.

12   Find the resultant of a force of $4\mathbf{i}$ at the origin together with a clockwise couple of moment 12.

13   Find the resultant of a force of $-6\mathbf{j}$ at the point $(2,0)$ together with an anticlockwise couple of moment 30.

14   Forces of $2\mathbf{j}$ and $-5\mathbf{j}$ act at the points $(4,0)$ and $(6,0)$, together with an anticlockwise couple of moment 10. Find the resultant.

15   Forces of $10\mathbf{i}$ and $6\mathbf{i}$ act at the points $(0, -2)$ and $(0,4)$, together with a clockwise couple of moment 20. Find the resultant.

16   Forces of $4\mathbf{j}$, $8\mathbf{j}$ and $F\mathbf{j}$ act at the points $(2,0)$, $(1,0)$ and $(-5,0)$. Find F if the resultant passes through the point $(-1,0)$.

17   Forces of $6\mathbf{j}$, $5\mathbf{j}$ and $F\mathbf{j}$ act at the points $(-2,0)$, $(3,0)$, $(5,0)$, together with a clockwise couple of moment 48. Find F if the resultant passes through the origin.

18   Forces of $-2\mathbf{i}$, $5\mathbf{i}$ and $F\mathbf{i}$ act at the points $(0,4)$, $(0, -2)$ and $(0,a)$. Find F and a, given that the resultant is an anticlockwise couple of moment 30.

Bodies in equilibrium under systems of parallel forces

Centre of gravity

The term 'centre of gravity' will be familiar to most readers. We shall deal with the topic in detail in a later chapter, but for the moment it will be sufficient to define the centre of gravity of a body as *the point at which all the body's weight can be considered to act*. The centres of gravity of uniform regular bodies are clearly at their geometrical centres.

Some worked examples will now be given involving bodies in equilibrium under parallel forces.

Example 1 A uniform rod AD of length 1 m and mass 5 kg rests horizontally on supports at B and C, where $AB = 15$ cm and $CD = 25$ cm. The rod carries a body of mass 2 kg at D (Fig. 6.9). Find (a) the forces on the supports, (b) the couple which will just make the rod tip about C and the new values of the forces on the supports when it is acting.

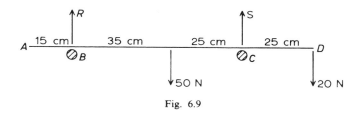

Fig. 6.9

Figure 6.9 shows the forces on the rod and not those on the supports. By Newton's third law, R and S are equal and opposite to the forces on the supports. Since the rod is uniform, its centre of gravity is at its mid-point.

(a) The rod is in equilibrium; hence there is no resultant moment about any point (principle of moments). We take moments about C to eliminate S and thus obtain an equation in which R is the only unknown:

$$60R + (20 \times 25) = 50 \times 25$$
$$\therefore\ R = 12\tfrac{1}{2}\ \text{N}.$$

Also since there is no resultant vertical force on the rod, we have

$$R + S = 70$$
$$\therefore\ S = 57\tfrac{1}{2}\ \text{N}.$$

(b) Let the required couple have an anticlockwise moment of L. When the rod is on the point of tipping about C the reaction at B must be zero. Hence, taking moments about C and remembering that a couple has the same moment about every point, we have

$$L + (50 \times 25) = 20 \times 25$$
$$\therefore\ L = -750.$$

The required couple is thus a **clockwise couple of moment 750 N cm.**
   Since the couple provides no extra vertical force, and the total downward force of 70 N has still to be balanced, the forces on the supports must now be **zero** and **70 N**.

Example 2 Bodies of weight $2W$ and $3W$ are attached at distances a and $5a$ from one end of a uniform beam of length $6a$ and weight W (Fig. 6.10). Find the position of a fulcrum on which the beam will balance in a horizontal position.

To balance the total downward force the fulcrum must exert an upward force of

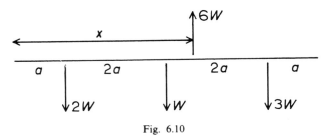

Fig. 6.10

$6W$ on the beam. Let it be a distance x from one end. Taking moments about the left end of the beam, we have

$$6Wx = 2Wa + 3Wa + 15Wa$$
$$= 20Wa$$
$$\therefore \ x = 3\tfrac{1}{3}a.$$

Example 3   A uniform beam AD of length $8a$ and weight W is suspended horizontally by two vertical strings at B and C where $AB = a$ and $CD = 2a$. When a certain weight is placed at A the tensions are in the ratio $3:2$. Where must the same weight be placed to equalise the tensions?

Let the unknown weight be w. When it is at A, let the tensions be $3T$ and $2T$, as shown in Fig. 6.11.

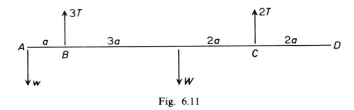

Fig. 6.11

The first task is to find w. Taking moments about B, we have

$$wa + 10Ta = 3Wa,$$

that is,

$$w + 10T = 3W. \quad (1)$$

Also since there is no resultant vertical force on the beam,

$$5T = W + w. \quad (2)$$

Eliminating T from (1) and (2):

$$w + 2(W + w) = 3W,$$

from which

$$w = W/3.$$

Now wherever we place the weight of $W/3$, there will be a total downward force of $4W/3$; hence if the tensions are equal they must each be $2W/3$. When this is so let the weight be a distance x to the right of B (Fig. 6.12).

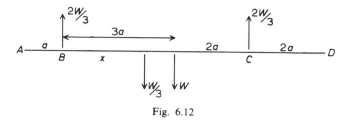

Fig. 6.12

Taking moments about B, we have

$$Wx/3 + 3Wa = 10Wa/3,$$

from which

$$x = a.$$

Example 4 Two uniform rods, ABC and CDE, of lengths $4a$ and $8a$, are freely jointed at C and rest horizontally on supports at B and D, where $BC = a$ and $DE = 2a$. The weight of rod ABC is W. Find (a) the weight of rod CDE, (b) the couple which must be applied to CDE to maintain equilibrium when a clockwise couple of $2Wa$ is applied to ABC.

(*Note*: In questions involving more than one movable object, it is essential when taking moments or resolving to state clearly which object or combination of objects is being referred to. Thus we write, for example, 'Taking moments *for the whole system*', 'Resolving vertically *for the rod AB*', etc.)

Letting the weight of rod CDE be w, the force diagram is as shown in Fig. 6.13. Note that there are in fact forces between the rods at C which we have not shown. These are internal to the whole system but external to each rod. Provided that we do nothing for each rod individually other than take moments about C, they can be ignored.

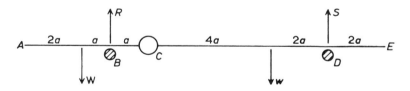

Fig. 6.13

(a) We begin by taking moments about C for each rod in turn:

Rod ABC: $Ra = 2Wa$, from which $R = 2W$.

Rod CDE: $6Sa = 4Wa$, from which $S = 2w/3$.

Now considering the vertical forces on the whole system (so that the internal forces at C cancel out), we have

$$W + w = 2W + 2w/3,$$

from which

$$w = 3W.$$

(b) The clockwise couple of $2Wa$ on ABC tends to move C downwards; so a *clockwise* couple on CDE will be required to maintain equilibrium. Let this couple have a moment of L. The reactions R and S will have altered, and we begin by finding their new values.

Taking moments about C for rod ABC,

$$2Wa = Ra + 2Wa,$$

from which

$$R = 0.$$

Now consideration of the vertical forces on the whole system gives $S = 4W$; hence, taking moments about C for rod CDE, we have

$$L + 12Wa = 24Wa$$

from which

$$L = 12\,Wa.$$

Exercise 6b

1   A uniform rod AB of length 40 cm and mass 7 kg rests horizontally on supports at A and a point 5 cm from B. Find the forces on the supports.

2   A uniform rod AD of length 60 cm and mass 6 kg rests horizontally on supports at B and C, where $AB = 10$ cm and $CD = 20$ cm. Find (a) the forces on the supports, (b) the vertical force at A which would just make the rod tip about B.

3   A uniform rod AB of length 1 m and mass 15 g carries bodies of mass 10 g and 45 g at A and B, respectively. Find the distance from A of the point about which it will balance:

4   A non-uniform rod AB of length 30 cm and weight 25 N rests horizontally on supports distant 2 cm from A and 8 cm from B. Given that the reactions of the supports are in the ratio $1:4$, find the distance of the centre of gravity from A.

5   A uniform rod AB of length 50 cm and weight 8 N rests horizontally on supports distant 10 cm from A and 5 cm from B. Find (a) the couples which just cause tipping about each of the supports, (b) the couple which equalises the forces on the supports.

6   A uniform rod AB of length $8a$ and weight $10W$ rests horizontally on supports at distances a and $6a$ from A. Find the forces on the supports. If in addition an anticlockwise couple of moment $5Wa$ acts on the rod, what are the new forces on the supports?

7   A non-uniform rod AD of length 1 m and mass 300 g rests horizontally on supports at B and C, where $AB = CD = 20$ cm. Given that the reaction at B is twice that at C, find the distance of the centre of gravity from A. Find also the distance from A at which a 200 g body would need to be placed to equalise the reactions.

8   Bodies of weight $2W$ and $4W$ are attached at distances a and $2a$ from the ends A and B of a uniform rod AB of length $10a$ and weight $2W$. Find the distance from A of the point about which the rod will balance. If in addition an anticlockwise couple of moment $12Wa$ is applied, by how much does this point shift?

9   A uniform rod AC of length 80 cm rests horizontally on supports at B, distant 16 cm from A, and another point P. Given that the ratio of the reaction at B to that at P is $2:3$, find the distance of P from C.

10   A uniform rod AB of length 1 m and mass 4 kg carries a 1 kg body at A and rests horizontally on supports distant 24 cm and 64 cm from A. Find (a) the forces on the supports, (b) the distance the rod must be moved to equalise these forces, (c) the couple which would have the same effect.

11 A non-uniform rod AB of length 120 cm balances about a point 42 cm from A when a 4 kg body is attached at A, and it balances about a point 30 cm from B when the body is moved to B. Find the mass of the rod and the distance of its centre of gravity from A.

12 Uniform rods AB, BC of lengths 40 cm and 60 cm are smoothly jointed at B and rest horizontally on supports distant 10 cm from A and x cm from C. The weights of AB and BC are respectively W N and 12 N, and the forces on the supports are respectively 6 N and R N. Find W, R and x.

13 Uniform rods AB, BC, of lengths $6a$, $4a$ and weights $4W$, W, respectively, are smoothly jointed at B and rest horizontally on supports distant $2a$ from A and x from C. Find the forces on the supports and the value of x in terms of a. If a body of weight $2W$ is attached at A, what couple must be applied to BC to maintain equilibrium?

14 Uniform rods AB, BC, of lengths $2a$, $4a$ and weights W, $2W$, respectively, are smoothly jointed at B and rest horizontally on supports distant x from A and a from C. Find the forces on the supports and the value of x in terms of a. If a couple of moment Wa, tending to move B downwards, is applied to BC, and a body of weight w, just sufficient to maintain equilibrium, is attached at A, find w and the new values of the forces on the supports.

7

General Systems of Coplanar Forces

Reduction of any system of coplanar forces

Consider a system of coplanar forces **A**, **B**, **C**, **D**, etc. We have seen in Chapters 4 and 6 that any pair from such a system, say **A** and **B**, is either in equilibrium or equivalent to a single resultant force or couple. Now let the resultant of **A** and **B** be combined with **C**. Since a couple together with a force is equivalent to an equal parallel force, there is certainly some single resultant of all three forces. This resultant can now be combined with **D**, and clearly the procedure can be continued until the whole system has been dealt with. The final result is one of the following three possibilities:

> **A single resultant force.**
> **A resultant couple.**
> **Equilibrium.**

The usual procedure for obtaining resultants involves a combination of the techniques discussed in Chapters 4 and 6. We begin by resolving all the forces into components in two convenient perpendicular directions. If the resultant is a force – as is the case unless both sums of components are zero – we can obtain its magnitude and direction from the two sums of components, and locate its line of action by taking moments about some convenient point. If the system either reduces to a couple or is in equilibrium, both the sums of components will be zero. In this case we take moments about one point only, thereby either obtaining the moment of the resultant couple or establishing that the system is in equilibrium.

The procedure will now be illustrated by worked examples.

Example 1 Forces of $4P$, P and $3P$ act along the sides AB, CB and CA of an equilateral triangle ABC of side a (Fig. 7.1). Find the magnitude of the resultant, the angle it makes with AB, and the point at which its line of action meets AB.

Here it is essential to resolve *along and perpendicular to AB*, in order to obtain the line of action in the required way. Letting the mid-point of AB be D, we have

$$\text{total component in direction } AB = 4P + P \cos 60° - 3P \cos 60°$$
$$= 3P.$$
$$\text{Total component in direction } CD = 3P \cos 30° + P \cos 30°$$
$$= 4P \cos 30°$$
$$= 2P\sqrt{3}.$$

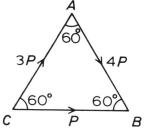

Fig. 7.1

Now we have the situation indicated in Fig. 7.2. By Pythagoras,

$$R^2 = 9P^2 + (4P^2 \times 3)$$
$$= 21P^2$$

Hence

$$R = 4.583P \quad \text{(or simply } P\sqrt{21}\text{)}.$$

Also

$$\tan \theta = 2P\sqrt{3}/3P,$$

where θ is the angle between the resultant and AB, from which

$$\theta = 49.11°.$$

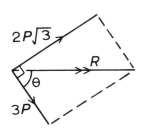

Fig. 7.2

Now let the line of action of the resultant cut AB at a distance x from A. At this point *resolve the resultant into its components* (Fig. 7.3). Using the principle

$$\text{moment of resultant} = \text{moment of original system},$$

we have, taking moments about A,

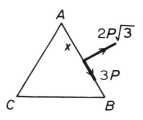

Fig. 7.3

where h is the height of the triangle; that is

$$2Px\sqrt{3} = Pa\frac{\sqrt{3}}{2} \qquad \text{(since } \sin 60° = \sqrt{3}/2)$$

$$\therefore x = a/4.$$

Example 2   Forces of $3i - 7j$, $6i$, $-7i + 3j$ act at the points whose position vectors are $2i + 4j$, $4i - 3j$, $-i + j$. Find the equation of the line of action of the resultant.

The position vectors are, of course, understood to specify points relative to the origin (see Chapter 3); hence the point with position vector $ai + bj$ is simply the point (a, b). The diagram is thus as shown in Fig. 7.4.

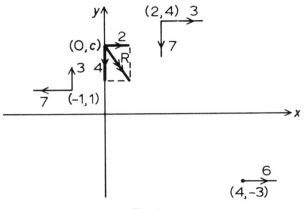

Fig. 7.4

The vector terminology means that the forces are already resolved into components, and we can therefore obtain the resultant in vector form simply by adding the coefficients of i and j:

$$R = 2i - 4j.$$

The gradient, m, of the line of action, is thus $-4/2 = -2$.

Now to find the equation of the line of action in the form $y = mx + c$, we need to locate the point $(0, c)$ at which the line of action meets the y axis. To do this we use once again the principle

moment of resultant = moment of original system.

Taking the origin, this gives

$$2c = (3 \times 4) + (7 \times 2) - (6 \times 3) - (7 \times 1) + (3 \times 1)$$
$$= 4.$$
$$\therefore c = 2$$

The equation of the line of action is thus $y = -2x + 2$.

Example 3   A force of 10 acts, in the upward direction, along the line $3x - 4y = 6$, together with forces of $-12i + 4j$ and $4i - 10j$ at the point $(-1, 2)$ and the origin, respectively. Show that this system reduces to a couple, and find its moment.

To sketch the graph of $3x - 4y = 6$, let y and x in turn be zero. This gives the points $(2, 0)$ and $(0, -1.5)$ and the graph is thus as shown in Fig. 7.5.

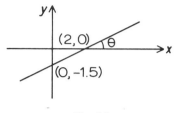

Fig. 7.5

The gradient, $\tan \theta$, is $\frac{3}{4}$, and hence $\cos \theta = \frac{4}{5}$ and $\sin \theta = \frac{3}{5}$. The x and y components of the force of 10 are thus 8 and 6, and if we resolve the force at the point where it meets the x-axis we have the complete force diagram shown in Fig. 7.6.

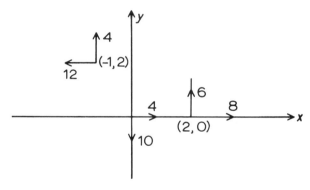

Fig. 7.6

Now since the sums of both the x and y components are zero, the system cannot reduce to a resultant force. It is therefore either equivalent to a couple or in equilibrium. To find out which, we need calculate the moment about one point only, and the most convenient point in this type of question is usually the origin.

$$\text{Total anticlockwise moment about the origin} = (6 \times 2) + (12 \times 2)$$
$$- (4 \times 1)$$
$$= 32.$$

The resultant of the system is thus **an anticlockwise couple of moment 32**.

Example 4 A force of P acts along each side of a regular hexagon $ABCDEF$ of side a, all the forces having the same sense (Fig. 7.7). Prove that the system is equivalent to a couple and find its moment.

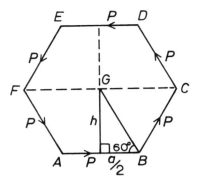

Fig. 7.7

By symmetry the sum of the components in any direction is zero. It follows that the system cannot reduce to a resultant force and must therefore either be in equilibrium or equivalent to a couple.

Total anticlockwise moment about the centre $G = 6Ph$

$$= 6P\frac{a}{2}\tan 60°$$

$$= 3Pa\sqrt{3} \quad \text{(since } \tan 60° = \sqrt{3}.\text{)}$$

The resultant is thus a couple, which is anticlockwise in the diagram drawn, of moment $3Pa\sqrt{3}$.

Example 5   Forces of 36 N, 15 N, 24 N, 10 N act along the sides AB, CB, DC, DA of a rectangle $ABCD$ in which $AB = 12$ cm and $BC = 5$ cm. A force of 65 N acts along the diagonal BD (Fig. 7.8). Prove that the system is in equilibrium.

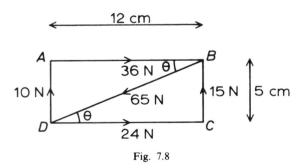

Fig. 7.8

Note first that since ABD is a 5, 12, 13 triangle, $\sin \theta = \frac{5}{13}$ and $\cos \theta = \frac{12}{13}$.

To prove that a system of coplanar forces is in equilibrium, we must show that it reduces neither to a resultant force nor to a resultant couple. To show that it does not reduce to a resultant force we show that the sums of the components in any two perpendicular directions are both zero; and having done this we prove that the system is not equivalent to a couple by showing that the moment about any one point is zero.

$$\text{Total component in direction } AB = 36 + 24 - 65 \cos \theta$$
$$= 60 - (65 \times 12/13)$$
$$= 0.$$
$$\text{Total component in direction } DA = 10 + 15 - 65 \sin \theta$$
$$= 25 - (65 \times 5/13)$$
$$= 0.$$

It follows that the system does not reduce to a resultant force. We now calculate the moment about any convenient point.

$$\text{Anticlockwise moment about } B = (24 \times 5) - (10 \times 12)$$
$$= 0.$$

The system therefore does not reduce to a couple. As it reduces neither to a resultant force nor to a resultant couple it must be in equilibrium.

Resultants of forces represented by directed line-segments

Theorem The resultant of forces completely represented by $m\overrightarrow{OA}$ and $n\overrightarrow{OB}$ is a force which is completely represented by $(m + n)\overrightarrow{OC}$, where C is the point which divides AB in the ratio $n:m$.

Proof Let C in Fig. 7.9 be the point which divides AB in the ratio $n:m$, i.e. the point such that $AC/CB = n/m$. Then we have

$$m\overrightarrow{OA} = m\overrightarrow{OC} + m\overrightarrow{CA} \qquad \text{(by the triangle law)},$$
$$n\overrightarrow{OB} = n\overrightarrow{OC} + n\overrightarrow{CB} \qquad \text{(by the triangle law)}.$$

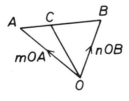

Fig. 7.9

Hence, adding,

$$m\overrightarrow{OA} + n\overrightarrow{OB} = (m + n)\overrightarrow{OC} + m\overrightarrow{CA} + n\overrightarrow{CB}.$$

Since, however, $AC/CB = n/m$, we have $mAC = nCB$, and thus the vectors $m\overrightarrow{CA}$ and $n\overrightarrow{CB}$ are equal and opposite. Hence

$$m\overrightarrow{OA} + n\overrightarrow{OB} = (m + n)\overrightarrow{OC}.$$

The line of action of the resultant passes through the point of intersection of the original two forces, i.e. at O; therefore the resultant is completely represented by $(m + n)\overrightarrow{OC}$ and the theorem is proved.

This theorem, together with the triangle theorem itself, can often be used to express the resultant of a system of forces as a directed line-segment when the original forces are expressed in this way. Some examples will now be given to illustrate this.

Example 1 G is the point of intersection of the medians of the triangle ABC, and D is any other point (Fig. 7.10). Prove that the resultant of the forces represented by $\overrightarrow{DA}$, $\overrightarrow{DB}$ and $\overrightarrow{DC}$ is completely represented by $3\overrightarrow{DG}$.

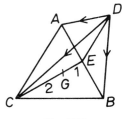

Fig. 7.10

We have

$$\overrightarrow{DA} + \overrightarrow{DB} = 2\overrightarrow{DE},$$

where D is the mid-point of AB, and

$$2\overrightarrow{DE} + \overrightarrow{DC} = 3\overrightarrow{DG},$$

where $CG/GE = 2:1$, and thus G is the point of intersection of the medians. Hence

$$\overrightarrow{DA} + \overrightarrow{DB} + \overrightarrow{DC} = 3\overrightarrow{DG}.$$

Since the original forces all act through D, the resultant also acts through D and thus is *completely* represented by $3\overrightarrow{DG}$.

Example 2   Forces completely represented by $\overrightarrow{AB}$, $2\overrightarrow{CB}$ and $\overrightarrow{CA}$ act along the sides of a triangle ABC (Fig. 7.11). Show that the resultant is represented in magnitude and direction by $3\overrightarrow{CB}$, and that its line of action trisects BA and CA.

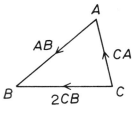

Fig. 7.11

By the triangle law, we have

$$\overrightarrow{AB} + \overrightarrow{CA} = \overrightarrow{CB},$$

acting at A, the point of intersection of $\overrightarrow{AB}$ and $\overrightarrow{CA}$. The system thus reduces to two parallel forces of $\overrightarrow{CB}$ and $2\overrightarrow{CB}$, and the resultant is therefore a force of $3\overrightarrow{CB}$, which divides any line from BC to A in the ratio $1:2$. The line of action thus trisects BA and CA.

Example 3 $ABCD$ is a quadrilateral in which E and F are the mid-points of AD and BC (Fig. 7.12). Prove that the resultant of the forces represented in magnitude and direction by $\overrightarrow{AB}$ and $\overrightarrow{DC}$ is represented in magnitude and direction by $2\overrightarrow{EF}$.

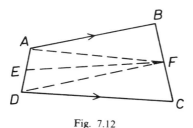

Fig. 7.12

We have
$$\overrightarrow{AB} = \overrightarrow{AF} + \overrightarrow{FB} \qquad \text{(by the triangle law)}$$
and
$$\overrightarrow{DC} = \overrightarrow{DF} + \overrightarrow{FC} \qquad \text{(by the triangle law)}.$$

Now since F is the mid-point of BC, $\overrightarrow{FB} = -\overrightarrow{FC}$. Hence, adding:
$$\overrightarrow{AB} + \overrightarrow{DC} = \overrightarrow{AF} + \overrightarrow{DF}$$
$$= 2\overrightarrow{EF} \qquad \text{by the theorem proved above.}$$

Note that the resultant is not (except in special cases) represented in line of action by $2\overrightarrow{EF}$.

Example 4 Forces completely represented by $6\overrightarrow{AB}$, $3\overrightarrow{CB}$ and $\overrightarrow{CA}$ act along the sides of a triangle ABC (Fig. 7.13). Find the resultant.

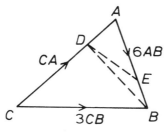

Fig. 7.13

We have $3\overrightarrow{CB} + 6\overrightarrow{AB} = 9\overrightarrow{DB}$, where D is the point such that $AD:DC = 3:6 = 1:2$, or $AD = \frac{1}{3}AC$.

In order to add the force of $\overrightarrow{CA}$ to that of $9\overrightarrow{DB}$, we re-express $\overrightarrow{CA}$ as $3\overrightarrow{DA}$. Then we can proceed:
$$3\overrightarrow{DA} + 9\overrightarrow{DB} = 12\overrightarrow{DE},$$

where E is the point such that $AE:EB = 9:3 = 3:1$, or $AE = \frac{3}{4}AB$. The resultant is thus a force which is completely represented by $\mathbf{12\overrightarrow{DE}}$, where D and E are defined above.

Miscellaneous problems on systems of forces

We have now completed our investigation of the determination of resultants, and proceed next to some more general problems on systems of coplanar forces. All of these will involve only the principles and theorems already considered.

Example 1   Figure 7.14 shows a rectangle in which $AB = 8$ units and $BC = 6$ units. Find P and Q if (a) the system of forces is equivalent to a couple, (b) the system reduces to a resultant force acting along BD. Show also that it is impossible for the system to be in equilibrium.

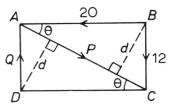

Fig. 7.14

(a) If the system is equivalent to a couple, the sum of the components in any direction is zero. Hence, resolving in the direction AB, we have

$$P \cos \theta = 20.$$

Now ABC is a $3, 4, 5$ triangle; so $\cos \theta = \frac{4}{5}$ and $\sin \theta = \frac{3}{5}$. The above equation thus becomes

$$P \times \tfrac{4}{5} = 20$$

Hence

$$P = 25.$$

Resolving now in the direction AD, we have

$$P \sin \theta + 12 = Q;$$

that is,

$$(25 \times 3/5) + 12 = Q$$

Hence

$$Q = 27.$$

(b) If the system reduces to a force along BD there must be no resultant moment about any point on this line. Since Q passes through D we can find P immediately by taking moments about this point:

$$Pd = (20 \times 6) - (12 \times 8)$$

that is,

$$P \times 8 \sin \theta = 24,$$

from which

$$P = 5.$$

Now, knowing P, we take moments about B to find Q:

$$Pd = Q \times 8$$

that is

$$5 \times 8 \sin \theta = 8Q$$

from which

$$Q = 3.$$

The system cannot be in equilibrium since, whatever the values of P and Q, there is certainly a moment about A of 12×8. A system in equilibrium must of course have no moment about any point.

Example 2 Forces of $4P$, $3P$ and $10P$ act along the sides BA, BC and CA of an equilateral triangle ABC of side a, and D is the mid-point of BC (Fig. 7.15). Prove that the system can be reduced to a force along DA together with a couple, and find the values of the force and the couple.

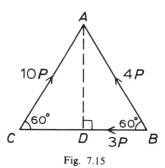

Fig. 7.15

We begin by resolving in the two perpendicular directions DA and DC.

$$\text{Total component in direction } DA = 10P \sin 60° + 4P \sin 60°$$
$$= 14P \sin 60° = 7P\sqrt{3};$$
$$\text{total component in direction } DC = 3P + 4P \cos 60° - 10P \cos 60°$$
$$= 3P + 2P - 5P = 0.$$

The resultant is thus a force of $\mathbf{7P\sqrt{3}}$ in the direction DA. Now whatever its actual line of action, the resultant is equivalent to an equal parallel force at any distance from itself, together with a couple. It must therefore be possible to express the system as a force of $7P\sqrt{3}$ along DA together with a couple.

To find the moment of the couple we use the fact that the moment of the new system about any point equals that of the original system. A convenient point to take is A, since the moment of the new system about this point is simply that of the couple. Hence we have

$$\text{Moment of couple} = \text{moment of original system about } A$$
$$= 3Pa \sin 60° \text{ clockwise}$$
$$= \mathbf{3Pa\sqrt{3}/2 \text{ clockwise}}.$$

Example 3 Forces of 12 N, 7 N, 6 N, 4 N act along the sides BA, BC, DC, DA of a square of side 6 m (Fig. 7.16). Find (a) the couple which must be added to the system to make the resultant act through B, (b) the additional force along DB which would make the resultant act through A.

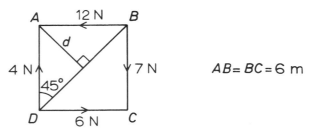

Fig. 7.16

(a) Let the couple have an anticlockwise moment of L N m. If the resultant acts through B there is no net moment about B. Hence

$$(6 \times 6) + L = 4 \times 6$$
$$\therefore \ L = -12.$$

The couple is thus **12 N m clockwise**.

(b) Let the required force along DB be F newtons. If the resultant acts through A there is no net moment about A. Hence

$$Fd = (7 \times 6) - (6 \times 6)$$

that is, $$F \times 6 \sin 45° = 6,$$

giving $$F = \sqrt{2}.$$

The force along DB is thus $\sqrt{2}$ **N**.

Example 4   Forces of $6\mathbf{i} - 5\mathbf{j}$ and $-4\mathbf{i} + 3\mathbf{j}$ act at the points with position vectors $-\mathbf{i} + 3\mathbf{j}$ and $2\mathbf{i} - \mathbf{j}$, respectively (Fig. 7.17). What force at the origin O reduces the system to a couple, and what is the couple's moment?

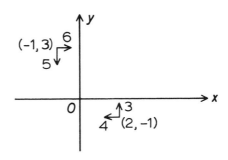

Fig. 7.17

The resultant $\mathbf{R}$ of the system is $2\mathbf{i} - 2\mathbf{j}$, and since a couple has a total component of zero in any direction the force which must be added at O is $-\mathbf{R}$, i.e. $-2\mathbf{i} + 2\mathbf{j}$. (This expression for the force specifies it completely if it is known to act at O; there is no need to calculate its magnitude and direction.)

To find the moment of the couple we note (a) that a couple has the same moment about every point, (b) that the force added to the system acts at O and therefore has no moment about O. It follows that the moment of the couple is simply that of the original system about O. That is:

anticlockwise moment of couple $= (3 \times 2) - (4 \times 1) + (5 \times 1) - (6 \times 3)$
$$= -11.$$

The couple thus has **a clockwise moment of 11**.

Example 5 The points A, B, C have position vectors $3\mathbf{i} + 2\mathbf{j}$, $2\mathbf{i} + 4\mathbf{j}$, and $5\mathbf{j}$, respectively. A certain force acts at O together with forces which are completely represented by $2\overrightarrow{AB}$ and $3\overrightarrow{BC}$. If the resultant cuts the axes at $(4, 0)$ and $(0, 2)$, find the force at O.

Denoting the position vectors of A and B by $\mathbf{a}$ and $\mathbf{b}$, and using the theorem for position vectors proved at the end of Chapter 3, we have

$$\overrightarrow{AB} = \mathbf{b} - \mathbf{a} = -\mathbf{i} + 2\mathbf{j}; \qquad \text{hence } 2\overrightarrow{AB} = -2\mathbf{i} + 4\mathbf{j};$$
$$\overrightarrow{BC} = \mathbf{c} - \mathbf{b} = -2\mathbf{i} + \mathbf{j}; \qquad \text{hence } 3\overrightarrow{BC} = -6\mathbf{i} + 3\mathbf{j}.$$

The forces can be resolved into their components at any points on their lines of action; hence, resolving the force of $2\overrightarrow{AB}$ at A and that of $3\overrightarrow{BC}$ at C, we have the complete force diagram shown in Fig. 7.18.

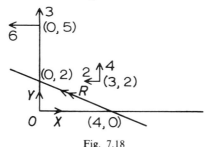

Fig. 7.18

The unknown force at O is shown as having components of X and Y. Now since the resultant passes through the points $(4, 0)$ and $(0, 2)$, the moment of the original system about each of these points is zero.

Taking moments about $(0, 2)$:

$$2X + (4 \times 3) + (6 \times 3) = 0$$
$$\therefore X = -15.$$

Taking moments about $(4, 0)$:

$$4Y + (4 \times 1) + (3 \times 4) = (2 \times 2) + (6 \times 5)$$
$$\text{i.e. } 4Y + 16 = 34$$
$$\therefore Y = 4\tfrac{1}{2}.$$

The force at O is therefore $-15\mathbf{i} + 4\tfrac{1}{2}\mathbf{j}$.

Exercise 7

1 Forces of P, $3P$, $4P$, $6P$ act along the sides AB, BC, DC, DA of a square $ABCD$ of side a. Find the magnitude of the resultant, the tangent of the angle it makes with CD, and the point at which its line of action cuts this line.

2   A triangle ABC has a right angle at B, and D is the foot of the perpendicular from B to AC. The side $AB = 6$ cm and $AC = 10$ cm. Forces of 15 N, 5 N, 8 N and 3 N act along AB, BC, CA and BD. Find the magnitude of the resultant and the point at which its line of action meets AC.

3   Forces of 6 N, 10 N and 4 N act along the sides AB, BC and CA of an equilateral triangle ABC of side 2 m. Find the magnitude of the resultant, the tangent of the angle it makes with AC, and the point at which its line of action meets AC.

4 $ABCD$ is a rectangle in which $AB = 8$ cm and $BC = 6$ cm. Forces of 12 N, 9 N, 3 N, 4 N, 20 N act along BA, BC, AD, CD, DB. Prove that the system is in equilibrium.

5 $ABCD$ is a rectangle in which $AB = 4$ m and $BC = 3$ m. Forces of 18 N, 8 N, 6 N, 10 N, 30 N act along AB, CB, DC, DA, BD. Prove that the system reduces to a couple and find its moment.

6 $ABCD$ is a rectangle in which E and F are the mid-points of AB and CD. $AB = 8$ cm and $BC = 3$ cm. Forces of 6 N, 5 N, 2 N, 10 N act along DC, FA, EF, FB. Find the magnitude of the resultant, the tangent of the angle it makes with AB, and the point at which its line of action meets this line.

7   Forces of $2i - j$, $-3i + 2j$, $4i - 7j$ act at the points whose position vectors are $i + 3j$, $-2j$, $-2i$. Find the equation of the line of action of the resultant. What force at O (in i, j form) will reduce the system to a couple and what is the couple's moment?

8   Equal forces of 10 N act along the sides AB, BC, CA of an equilateral triangle ABC of side 4 m. Prove that the system is equivalent to a couple and find its moment.

9   Forces which are completely represented by $\overrightarrow{AB}$, $\overrightarrow{BC}$ and $3\overrightarrow{AC}$ act along the sides of a triangle ABC. Specify the resultant completely.

10 $ABCD$ is a rectangle in which $AB = 12$ m and $BC = 6$ m. Forces of 10 N, 14 N, 8 N, 4 N, 26 N act along BA, CB, CD, DA, AC. Find the magnitude of the resultant and the point at which its line of action meets BC.

11   Forces of $6F$ and $4F$ act along the sides BC and AC of an equilateral triangle ABC of side $2a$. Find the point at which the line of action of the resultant meets AB. What couple must be added to the system to make the resultant act through B?

12   Forces of $2P$, $4P$, $6P$ act along the sides AB, BC, AC of an equilateral triangle ABC of side a. Find the magnitude of the resultant and deduce the perpendicular distance from B of its line of action.

13 $ABCD$ is a parallelogram. Forces act at a point which are represented in magnitude and direction (but not line of action) by $\overrightarrow{BA}$, $\overrightarrow{BC}$, $3\overrightarrow{DC}$ and $\overrightarrow{AD}$. Prove that the resultant is represented in magnitude and direction by $2\overrightarrow{AC}$.

14   Forces of 10, P, Q, S act along the sides AB, BC, CD, DA of a rectangle $ABCD$ in which $AB = 5$ and $BC = 2$. Find P, Q and S if the system is in equilibrium.

15   Forces of $3i - 5j$ and $2i + j$ act at the points $(-3, 4)$ and $(5, -1)$, respectively. A clockwise couple of moment 6 also acts. Find the magnitude of the resultant, the tangent of the angle it makes with the positive x-axis (taking anti-clockwise as positive), and the point at which its line of action cuts the x-axis.

16 $ABCD$ is a quadrilateral in which the mid-points of AC and BD are P and Q. Prove that the resultant of the forces which are completely represented by $\overrightarrow{AB}$, $\overrightarrow{CB}$, $\overrightarrow{CD}$, $\overrightarrow{AD}$ is completely represented by $4\overrightarrow{PQ}$.

17 ABC is a triangle in which $AC = 4$ m, $\angle ACB = 90°$ and $\angle BAC = 60°$. The point D is the foot of the perpendicular from C to AB. Forces of 2 N, 8 N and $4\sqrt{3}$ N act along BA, CA and CB. Prove that the system is equivalent to a force along CD together with a couple, and find the magnitudes of the force and the couple.

18 A force of 4 N acts along each of the sides of a regular pentagon of side 2 m, each force having the same sense, relative to the pentagon. Prove that the system is equivalent to a couple and find its moment.

19 $ABCD$ is a rectangle in which $AB = 24$ and $BC = 7$. Forces of $P, 28, 12, 14, Q$ act along AB, CB, CD, AD, BD. Find P and Q if (a) the system is equivalent to a couple, (b) the resultant acts along CD.

20 The medians of triangle ABC meet at G. Prove that the system of forces represented by $\overrightarrow{GA}, \overrightarrow{GB}, \overrightarrow{GC}$ is in equilibrium.

21 $ABCDEF$ is a regular hexagon of side a. Forces of $2P, 5P, 4P$ act along BC, DE, EB. Find the point at which the line of action of the resultant meets AB.

22 A force of $3\mathbf{i} - 6\mathbf{j}$ acts at the point $(4, 3)$, together with an anticlockwise couple of moment 9. Find the equation of the line of action of the resultant.

23 $ABCD$ is a quadrilateral in which E, F, G, H are the mid-points of AB, BC, CD, DA, and O is any point in the plane of the quadrilateral. Prove that the system of forces represented by $\overrightarrow{OA}, \overrightarrow{OB}, \overrightarrow{OC}, \overrightarrow{OD}$ is equivalent to that represented by $\overrightarrow{OE}, \overrightarrow{OF}, \overrightarrow{OG}, \overrightarrow{OH}$.

24 Points A, B, C have position vectors $2\mathbf{i} - \mathbf{j}, \mathbf{i} + 3\mathbf{j}, -\mathbf{i} + 2\mathbf{j}$. Express in terms of $\mathbf{i}$ and $\mathbf{j}$ the resultant of the forces which are completely represented by $2\overrightarrow{AB}$ and $3\overrightarrow{CB}$, and find the point at which the line of action of the resultant cuts the x-axis.

25 A force of magnitude 10 acts, in the upward direction, along the line whose equation is $3x - 4y = 12$. Also a force of $2\mathbf{i} - \mathbf{j}$ acts at the point $(2, 1)$, together with a couple. Given that the resultant acts through the point $(1, 0)$, find the couple and an expression for the resultant in terms of $\mathbf{i}$ and $\mathbf{j}$.

26 A certain force has a component of 2 in the direction of the positive x-axis, and its anticlockwise moments about the origin and the point $(3, 0)$ are respectively -4 and 6. Find the equation of its line of action.

27 ABC is an equilateral triangle, D being the mid-point of BC. Forces of $P, 2, 4$, Q act along BA, CA, CB, AD. Find P and Q if (a) the system is equivalent to a couple, (b) the resultant acts along BC. Explain why the system cannot be in equilibrium.

28 $ABCD$ is a rectangle in which $AB = 12$ m and $BC = 5$ m. Forces of 20 N, 15 N, 40 N, 20 N, 65 N act along AB, BC, DC, AD, CA. Show that the system can be reduced to a force along AD together with a couple, and find the magnitudes of the force and the couple.

29 A force of 20 acts along the line OA, where A is the point $(3, 4)$. Also a force of $4\mathbf{j}$ acts along the y-axis. Find (a) the couple which must be added to the system to make the resultant act through $(2, 0)$, (b) the force required at O to make the resultant $-3\mathbf{i}$.

30 $ABCD$ is a trapezium in which $AB = BC = AD = a$, and $\angle C = \angle D = 60°$. Forces of $P, 2P, 4P, 4P$ act along the sides AB, BC, CD, DA. Show that the resultant is perpendicular to CD and find the point at which its line of action cuts this line.

8

Impulse, Momentum, Impact

Impulsive forces

Consider any sharp *blow*, such as that produced by a hammer on striking a nail, or that occurring during the collision of two billiard balls. The two objects concerned in such a blow are in contact for a very short time, but the force which each exerts on the other during that time is very great. This can be seen by the fact that the bodies undergo virtually instantaneous changes in velocity. Forces of the kind described are called *impulsive*, and impulsive forces may be defined generally as ones which are large enough to cause significant changes in the velocities of the bodies they act upon, yet brief enough to have negligible effects on the positions of those bodies during the periods for which they act.

Besides being very large, impulsive forces vary rapidly over the short times for which they act, and it is therefore clear that they themselves cannot be physically measured with any ease. The measurement of the size of a blow therefore presents a problem. A solution is suggested by the graph of an impulsive force against time, which in general has the form of Fig. 8.1.

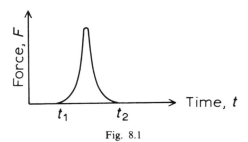

Fig. 8.1

Consider the *area* between the graph and the time axis. This area can be shown, with the help of a little calculus, to represent a quantity which is physically measurable. For the derivation, note that we need to use the strict form of Newton's second law, namely $F = \mathrm{d}(mv)/\mathrm{d}t$, rather than the simple $F = ma$. We have

$$\text{area below graph} = \int_{t_1}^{t_2} F \, \mathrm{d}t$$

$$= \int_{t_1}^{t_2} \frac{\mathrm{d}(mv)}{\mathrm{d}t} \, \mathrm{d}t$$

$$= \int_{(mv)_1}^{(mv)_2} d(mv) \qquad \text{(where } (mv)_1 \text{ and } (mv)_2 \text{ are the initial}$$
and final values of the momentum)

= final momentum − initial momentum

= total change in momentum.

Now this quantity, the total change in momentum caused by a blow, is physically measurable, and therefore suitable for use as a measure of the size of a blow. It is known as the *impulse* of the force between the two bodies concerned.

The idea of impulse need not be limited in its application to sharp blows in which there are sudden changes of velocity; the above derivation is applicable generally and shows that for any force acting for any period of time we have

> **Impulse of a force $F = \int F \, dt$**
>
> $\qquad$ **= total change in momentum caused by F.**

The SI unit of impulse has two alternative forms. The unit can be derived from the definition and written N s, or the unit of momentum, $kg \, m \, s^{-1}$, can be used. The two forms are equivalent since $1 \, N = 1 \, kg \, m \, s^{-2}$.

Example Suppose a ball of mass 200 g strikes a wall at $20 \, m \, s^{-1}$ and rebounds at $15 \, m \, s^{-1}$. To find the impulse received by the ball we proceed thus.

$\qquad$ Total change of velocity in direction of impulse $= 35 \, m \, s^{-1}$.

$\qquad$ Hence total change in momentum $= 0.2 \times 35$

$$= 7 \, kg \, m \, s^{-1}.$$

$\qquad$ Hence impulse received $= \mathbf{7 \, N \, s} \quad$ or $\quad \mathbf{7 \, kg \, m \, s^{-1}}$.

The principle of conservation of (linear) momentum

This principle may be stated as follows.

> **The total (linear) momentum of a system of bodies is constant in any direction in which no external forces act.**

The proof of the principle is simple. Since no external forces act, the momentum of any individual body can be altered only by its internal interaction with another body in the system. (This interaction may take various forms, e.g. a collision, gravitational attraction, magnetic attraction and repulsion, etc.) Now by Newton's third law, whenever two bodies interact the forces which each exerts on the other are equal and opposite. Also these forces obviously act for the same times. It follows that the impulses, and hence the momenta, which the two bodies receive are equal and opposite, and thus that the total momentum of the system is unchanged.

Care must be taken over the *directions* in which the principle is applied. In the case of collisions, etc., the principle holds for any direction in which no external *impulsive* forces act. Any non-impulsive forces can be ignored in such cases, since they have no significant effect in the very short time for which impulsive forces act. The importance of direction is illustrated in the first of the following examples.

Example 1   A 200 kg gun, mounted on wheels, stands on a smooth plane inclined at 30° to the horizontal. It fires a 50 g bullet horizontally at a speed of 600 m s^{-1}, and itself recoils straight up the plane (i.e. along a line of greatest slope) (Fig. 8.2). Find the initial speed of the gun.

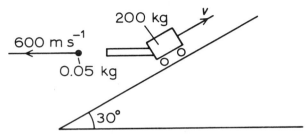

Fig. 8.2

Since the impulsive force between the bullet and the gun has a component perpendicular to the slope, the ground applies a reactive impulse which is external to the system. In the absence of friction this reaction is normal to the slope; hence we can apply the principle of conservation of momentum *along* the slope. This is the *only* direction (in the plane of the paper) in which the principle can be applied, for the external impulse has a component in every other direction. We therefore proceed as follows.

By the principle of conservation of momentum, applied along a line of greatest slope, we have

$$0.05 \times 600 \cos 30° = 200v$$
$$\therefore \quad v = 0.130 \text{ m s}^{-1}.$$

Note that the force of gravity, which does have a component along the slope, can be ignored when we are dealing with the action between the bullet and the gun since this action is impulsive while the force of gravity is not.

Example 2   A bullet of mass 20 g is fired downwards at an angle of 20° to the horizontal into a block of mass 3 kg standing at rest on a smooth horizontal surface (Fig. 8.3). Given that the bullet becomes embedded in the block and their common velocity is then 2.5 m s^{-1}, find the original velocity of the bullet.

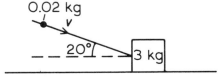

Fig. 8.3

Here the external reaction of the ground is vertical; hence the horizontal momentum is conserved and we have

$$0.02v \cos 20° = 3.02 \times 2.5$$
$$\therefore v = \mathbf{401.7\ m\,s^{-1}}.$$

Example 3   A piledriver of mass 6 tonnes falls from a height of 4 m onto a pile of mass 2 tonnes. If the average frictional resistance of the ground is 2×10^6 N, find the distance the pile penetrates.

First we find the velocity, v_1, of the driver just before it strikes the pile. Using

$$v^2 = u^2 + 2as,$$

we have

$$v_1{}^2 = 2 \times 10 \times 4 = 80,$$

and hence

$$v_1 = 8.944\ m\ s^{-1}.$$

Now letting the common velocity of pile and driver just after their collision be v_2, we have, by the principle of conservation of momentum,

$$6 \times 8.944 = 8v_2$$
$$\therefore v_2 = 6.708\ m\ s^{-1}.$$

Finally, to find the distance penetrated – s, say – we can either use the principle of conservation of energy or find the deceleration and then use a kinematics equation. Adopting the latter method we proceed as follows.

The pile and driver can now be regarded as a single body of mass 8 tonnes. Hence we have

$$
\begin{aligned}
\text{resultant upward force} &= \text{resistance} - \text{weight} \\
&= 2 \times 10^6 - 80\,000 \\
&= 1\,920\,000\ \text{N}.
\end{aligned}
$$
$$
\begin{aligned}
\therefore \text{deceleration} &= \text{resultant force/mass} \\
&= 1\,920\,000/8\,000 \\
&= 240\ \text{m s}^{-2}.
\end{aligned}
$$

Now using $v^2 = u^2 + 2as$, with $v = 0$, $u = 6.708$ and $a = -240$, we have

$$0 = 6.708^2 - (2 \times 240 \times s)$$

from which

$$s = 0.09374\ m \quad \text{or about} \quad \mathbf{9.4\ cm}.$$

It should be realised that while momentum is conserved, there is always a loss of KE at a collision between ordinary objects. The final example illustrates this.

Example 4   A 10 kg block moving at 4 m s^{-1} strikes a 15 kg block which is at rest, and the two blocks then move together. Find the loss of KE at the impact.

Letting the common velocity be v and using the principle of conservation of momentum, we have

$$10 \times 4 = 25v$$
$$\therefore v = 1.6\ m\ s^{-1}.$$

Now using the equation $KE = \frac{1}{2}mv^2$, we obtain

$$KE \text{ before impact} = \frac{1}{2} \times 10 \times 4^2$$
$$= 80 \text{ J}$$
$$KE \text{ after impact} = \frac{1}{2} \times 25 \times 1.6^2$$
$$= 32 \text{ J}.$$

There is thus a loss of KE at the impact of **48 J**.

Exercise 8a

1 A 2 kg ball strikes a wall at right angles at 4 m s^{-1}. Find the impulse on the wall if (a) the ball is brought to rest, (b) it rebounds at 3 m s^{-1}.

2 A 3 kg body is dropped from a height of 5 m. Find the impulse on the ground if (a) the body is brought to rest, (b) it rebounds to a height of 3 m.

3 Find the impulse of a constant force of 5 N, acting for 6 s. By how much would this impulse increase the speed of a 3 kg body?

4 A 4 kg block moving at 5 m s^{-1} strikes a 6 kg block which is at rest. If the blocks stay together after impact find their common velocity and the loss of KE at the impact.

5 A 10 kg body strikes a 4 kg body which is at rest, and the two then move together with a speed of 20 m s^{-1}. Find the original speed of the 10 kg body and the impulse at the collision.

6 A 2 kg body is dropped from rest at a height of 60 m. After 2 s it meets a 3 kg body moving upwards at 10 m s^{-1}, and the two bodies then move together. Find the height of the pair of bodies after a further second.

7 A body of mass m moving with a speed of v meets a body of mass M, initially at rest, and the two then move together. Obtain an expression for the loss of KE at the collision.

8 A bullet of mass 25 g is fired downwards at an angle of $30°$ to the horizontal into a 2 kg block standing on smooth horizontal ground. Given that the original speed of the bullet is 500 m s^{-1}, find the common velocity of bullet and block and the impulse imparted to the block.

9 A 500 g hammer, moving horizontally at 12 m s^{-1}, strikes a 20 g nail and rebounds off it at 4 m s^{-1}. Find the impulse received by the nail and its initial velocity. Given that it moves 0.5 cm into a piece of wood, find the average resistance of the wood.

10 A piledriver of mass 8 tonnes falls from a height of 5 m onto a pile of mass 2 tonnes. Given that the average frictional resistance of the ground is 10^6 N, find the distance penetrated.

11 A piledriver of mass 4 tonnes falls from a height of 180 cm onto a pile of mass 2 tonnes, driving it 10 cm into the ground. Find the average resistance of the ground.

12 A 500 kg gun standing on smooth level ground fires a 20 g bullet at an angle of $15°$ to the horizontal at a speed of 400 m s^{-1}. Find the velocity of recoil of the gun.

13 A 100 kg gun stands on a smooth plane of inclination $20°$. It fires a 30 g bullet horizontally and recoils straight up the plane at a speed of 20 cm s^{-1}. Find the velocity of the bullet.

14 A 20 kg body, initially at rest, splits into two pieces with masses of 12 kg and 8 kg as a result of an internal explosion. If the smaller piece flies off at 120 m s^{-1} find the speed at which the other piece does so. Find also the KE supplied by the explosion.

15 A 6 kg body, initially at rest, explodes into two pieces with masses 4 kg and 2 kg which fly apart. If the KE supplied by the explosion is 2.4×10^5 J, find their velocities.

16 A body of mass m is moving with a speed of v. As a result of an internal explosion it splits into two parts with masses in the ratio $3:1$, the larger of which is brought to rest. Prove that the KE supplied by the explosion is $3mv^2/2$.

Newton's law of impact

As the above examples show, the principle of conservation of momentum is sufficient to solve collision problems in which no *rebounding* occurs, so that the two colliding bodies move together after impact. In general however bodies separate after impact, and to deal with such cases we need a further law which was discovered experimentally by Newton.

Consider first the simplest case – that of a moving body striking a fixed one, e.g. a ball striking a wall. Here experiment shows that for any definite pair of bodies *the velocity just after the collision is directly proportional to the velocity just before it*. Expressing this as an equation, we have

$$v = ku,$$

where v and u are respectively the velocities after and before the collision.

Now v and u are *signed* quantities, being positive or negative according to their directions. Since there is necessarily a change of direction at a collision between a moving body and a fixed one, v and u must in fact be opposite in sign, and thus k in the above equation must be negative. To avoid this, and obtain a positive constant of proportionality, a minus sign is introduced into the equation, and the new, positive constant is called e. This gives us

$$v = -eu.$$

Let us now consider what e measures. The greater the value of e, the greater is v for a given value of u. Thus e measures the elasticity or 'bounciness' of the two bodies, and consequently is known as their *coefficient of restitution*. Clearly e cannot be greater than 1. If $e = 1$ (which is true for molecules but impossible for ordinary bodies), the bodies are said to be *perfectly elastic*, while if $e = 0$ (so that the bodies do not separate after colliding) they are described as *inelastic*. Unless $e = 1$ there is necessarily a *loss of energy* at a collision, the energy being mainly converted to heat.

Suppose now that both colliding bodies are free to move, and consider a direct impact in which the velocities before and after the collision are as shown in Fig. 8.4.

(The u's and v's in Fig. 8.4 can of course be negative, indicating movement in the opposite direction to that shown; but diagrams of this kind should always be drawn with all the arrows in the same direction.)

Now the quantity here which corresponds to the u of the body striking a fixed wall is $u_1 - u_2$, i.e. the *overtaking* velocity or *relative* velocity. After the collision the

Before:

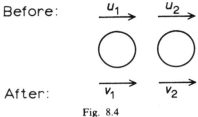

After:

Fig. 8.4

relative velocity is $v_2 - v_1$ and the law thus becomes

$$v_2 - v_1 = e(u_1 - u_2),$$

that is,

$$v_1 - v_2 = -e(u_1 - u_2)$$

This is Newton's law of impact in its general form. In words it may be stated as follows:

If two bodies collide directly their relative velocity just after the collision is directly proportional to their relative velocity just before it.

In this book we shall deal only with *direct* collisions, i.e. those in which the velocities are perpendicular to the surface which is common to the two bodies at impact. In the case of two balls, this means that the velocities are along the line joining the centres. Oblique collisions are handled by resolving along and perpendicular to the common surface. Newton's law continues to hold for the perpendicular components, and, in the absence of friction, the tangential components are unaffected by the collision.

Problems on impact are solved by using Newton's law together with the principle of conservation of momentum to obtain two simultaneous equations. The procedure in some typical problems will now be illustrated by worked examples.

Worked examples on collisions

The following points should be noted regarding the presentation of problems involving collisions.

1. It is not necessary to draw a diagram showing a collision between a moving body and a fixed one. Here Newton's law effectively states simply that the *speed* after the collision is e times the speed before it.
2. A separate diagram should be drawn for each individual collision between movable bodies. Example 1 below shows a convenient format for such diagrams. Note in particular that *all arrows should be in the same direction*, in each diagram.
3. Known values should be written straight into the diagram, letters being used only for unknown quantities.

Example 1 A ball of mass 2 kg, moving at 4 m s^{-1}, collides directly with a ball of mass 3 kg which is moving in the opposite direction at 5 m s^{-1} (Fig. 8.5). Given that $e = \frac{1}{2}$, find (a) the velocities after the impact, (b) the impulse between the balls, (c) the loss of KE.

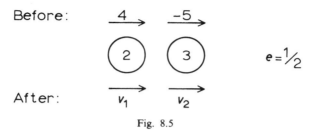

Fig. 8.5

(a) By the principle of conservation of momentum,

$$2v_1 + 3v_2 = 8 - 15,$$

that is,

$$2v_1 + 3v_2 = -7. \quad (1)$$

By Newton's law,

$$v_1 - v_2 = -e(u_1 - u_2),$$

that is,

$$v_1 - v_2 = -\tfrac{1}{2}(4 + 5),$$

that is,

$$2v_1 - 2v_2 = -9. \quad (2)$$

$(1) - (2)$:

$$5v_2 = 2$$

$$\therefore \ v_2 = 0.4 \text{ m s}^{-1}.$$

Substituting 0.4 for v_2 in (2):

$$2v_1 - 0.8 = -9$$

from which

$$v_1 = -4.1 \text{ m s}^{-1}.$$

(b) The impulse each ball receives is equal to its change in momentum. Taking the 2 kg ball we have

Momentum before impact $= 2 \times 4 = 8 \text{ kg m s}^{-1}$ to the right.
Momentum after impact $= 2 \times 4.1 = 8.2 \text{ kg m s}^{-1}$ to the left.
Hence impulse $= 16.2 \text{ kg m s}^{-1}$ or **16.2 N s.**

(c) Using the equation $\text{KE} = \frac{1}{2}mv^2$, we have

$$\text{Total KE before impact} = (\tfrac{1}{2} \times 2 \times 4^2) + (\tfrac{1}{2} \times 3 \times 5^2)$$

$$= 53.5 \text{J}.$$

$$\text{Total KE after impact} = (\tfrac{1}{2} \times 2 \times 4.1^2) + (\tfrac{1}{2} \times 3 \times 0.4^2)$$

$$= 17.05 \text{ J}.$$

$$\text{Hence loss of KE} = \textbf{36.45 J}.$$

Example 2 Balls *A* and *B*, with masses in the ratio 2 to 1, lie on a smooth horizontal surface. *A* is set in motion and collides directly with *B*, which goes on to impinge directly on a vertical wall. It rebounds and strikes *A* again, thus bringing *A* to rest. Given that the coefficient of restitution of the two balls is $\frac{2}{3}$, find that of *B* and the wall.

Let the masses be 2 units and 1 unit. (This makes cancellation by m unnecessary.)
Let the initial velocity of A be u and the required coefficient be e.

First collision   (Fig. 8.6)

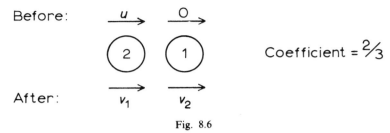

Fig. 8.6

By the principle of conservation of momentum,
$$2v_1 + v_2 = 2u. \qquad (1)$$
By Newton's law,
$$v_1 - v_2 = -\tfrac{2}{3}u. \qquad (2)$$
$(1) + (2)$:
$$3v_1 = 4u/3$$
$$\therefore \; v_1 = 4u/9.$$
Substituting $4u/9$ for v_1 in (1):
$$8u/9 + v_2 = 2u$$
$$\therefore \; v_2 = 10u/9.$$

B now strikes the wall and rebounds with a speed of $10ue/9$, after which it collides with A again, bringing the latter to rest. Consider this collision.

Third collision   (Fig. 8.7)

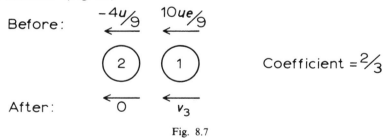

Fig. 8.7

By the principle of conservation of momentum,
$$v_3 = -8u/9 + 10ue/9. \qquad (3)$$
By Newton's law,
$$v_3 = -\tfrac{2}{3}(10ue/9 + 4u/9). \qquad (4)$$
Eliminating v_3 from (3) and (4):
$$-8u/9 + 10ue/9 = -\tfrac{2}{3}(10ue/9 + 4u/9)$$
$$\therefore \; -24 + 30e = -20e - 8 \quad \text{(multiplying both sides by } 27/u)$$
$$\therefore \; 50e = 16$$
$$\therefore \; e = 8/25.$$

Example 3   A ball is dropped from a height of 5 m onto a horizontal plane. $e = 0.6$. Find the time it takes for the ball finally to come to rest.

First we find the time t that the ball takes to reach the plane, and the velocity v with which it strikes it. Using

$$s = ut + \tfrac{1}{2}at^2,$$

we have

$$5 = 5t^2$$

$$\therefore \ t = 1 \text{ s.}$$

Also from

$$v = u + at,$$

we have

$$v = 10 \times 1$$

$$= 10 \text{ m s}^{-1}.$$

The ball rebounds with a speed of $10 \times 0.6 = 6$ m s^{-1}. Now we find the time it takes to return to the plane. Since

$$u = 6,$$
$$a = -10,$$
$$s = 0,$$

we have, using

$$s = ut + \tfrac{1}{2}at^2,$$
$$0 = 6t - 5t^2,$$
$$\therefore \ t = 1.2 \text{ s.}$$

The ball returns to the plane with a speed of 6 m s^{-1} once more, and rebounds from it with a speed of 0.6×6 m s^{-1}.

By repeating the calculation just carried out, but making u equal to 0.6×6, we can easily show that the time the ball takes to return yet again is 0.6×1.2 s. Similarly the time it takes for its next flight is $0.6^2 \times 1.2$ s, and the expression for the total time the ball is in motion is

$$1 + 1.2 + (0.6 \times 1.2) + (0.6^2 \times 1.2) + \cdots$$
$$= 1 + 1.2(1 + 0.6 + 0.6^2 + \ldots)$$

Now the sum to infinity of the G.P. $a + ar + ar^2 + \ldots$ is $a/(1 - r)$. Hence the total time is

$$1 + 1.2 \times \frac{1}{1 - 0.6}$$

$$= 4 \text{ s.}$$

Exercise 8b

(Coefficients of restitution are denoted by e.)

1   A 4 kg ball, moving horizontally, strikes a vertical wall at 12 m s^{-1} and rebounds at 9 m s^{-1}. Find e for the ball and the wall, and the KE lost at the impact.
2   A ball is dropped from a height of 135 cm onto horizontal ground and rebounds to a height of 60 cm. Find e.

3 A ball dropped onto horizontal ground takes $1\frac{1}{4}$ s to reach the ground and another $\frac{3}{4}$ s to come to instantaneous rest once more. Find e.

4 A ball is dropped from a height of 16 m onto horizontal ground. Given that $e = \frac{5}{8}$, find the height to which it rises.

5 A ball of mass 4 kg, moving at 2 m s^{-1}, collides directly with a stationary 3 kg ball. $e = \frac{3}{4}$. Find the speeds after the collision and the loss of KE. at the collision.

6 A 3 kg ball moving at 10 m s^{-1} collides directly with a 2 kg ball which is moving in the same direction at 5 m s^{-1}. $e = \frac{1}{2}$. Find the speeds after the collision and the impulse at the collision.

7 A 2 kg ball moving at 4 m s^{-1} collides directly with a 3 kg ball which is moving in the opposite direction at 2 m s^{-1}. $e = \frac{1}{3}$. Find the speeds after the collision and the loss of KE at the collision.

8 Two perfectly elastic balls, with masses of 1 kg and 4 kg, are moving in opposite directions with speeds of 15 m s^{-1} and 5 m s^{-1}, respectively. Find their speeds after the collision and the impulse at the collision.

9 A 2 kg ball strikes a stationary 3 kg ball and is itself brought to rest. Find e.

10 Prove that perfectly elastic balls of equal mass exchange velocities when colliding directly.

11 A ball of mass m, moving with a speed of u, collides directly with a ball of mass $3m$ and rebounds with a speed of $2u$. $e = \frac{1}{4}$. Find the original speed of the second ball.

12 A body A, moving due east with a speed of $5u$, collides directly with an identical body B, and after the collision B moves due east at a speed of $4u$. Given that $e = \frac{5}{6}$, find the velocity of B before the collision and that of A after the collision.

13 A ball of mass m collides directly with one of mass $2m$ which is initially at rest. After the collision the balls move in the same direction with speeds in the ratio $1:2$. Find e.

14 A ball moving with a speed of $10u$ collides directly with an identical ball which is moving in the same direction at $4u$. After the collision the relative speed is $2u$. Find the actual speeds after the collision and the value of e.

15 Two balls A and B, with masses in the ratio $2:1$, lie at rest on a smooth horizontal surface. A is projected towards B at a speed of u, and after their collision B rebounds off a wall and directly strikes A again. Given that $e = 1$ for each collision, find A's final speed.

16 Three identical balls A, B, C lie in a straight line on a smooth horizontal surface. $e = \frac{1}{2}$ for each pair of balls. A is projected towards B with a speed of u. Determine the total number of impacts and the final speeds of the balls.

17 Balls A, B, C, of masses m, $2m$, m, lie at rest in a straight line on a smooth horizontal surface. $e = \frac{1}{3}$ for each pair of balls. A is projected towards B with a speed of u. Determine the total number of collisions and the final speeds.

18 Two identical particles A and B lie at the opposite ends of a diameter of a smooth horizontal circular groove of circumference 4 m. For collisions between A and B, $e = \frac{1}{2}$. If A is given a speed of 2 m s^{-1}, what time will elapse before the second collision between A and B occurs?

19 Two balls A and B, with masses in the ratio $1:2$, lie at rest on a smooth horizontal surface. A is projected towards B with a speed of 10 m s^{-1}, and after their collision B travels 12 m to a vertical wall, rebounds, and strikes A directly again. Given that $e = 0.2$ for the two balls, and 0.25 for B and the wall, find the time interval between the two collisions of A and B.

20 Three balls A, B, C, of equal mass, lie at rest in a straight line on a smooth horizontal surface. The coefficient of restitution for A and B is e. Find in terms of e the maximum possible value of the coefficient for B and C if only two collisions occur when A is projected towards B.

21 Balls A, B, C, with masses of m, $2m$, $3m$, lie at rest in a straight line on a smooth horizontal surface. e is the same for each pair of balls. Show that, when A is projected towards B, the condition for more than two collisions to occur is that $e^2 - 3e + 1 > 0$. Solve this inequality, to 2 d.p.

22 Two identical balls A and B lie at rest on a smooth horizontal surface. A is projected towards B, and after their collision B strikes a vertical wall, rebounds, and strikes A again. As a result, A moves away from the wall at half its original speed. Given that e for the two balls is $\frac{3}{4}$, find e for B and the wall.

23 Two light inextensible strings of equal length are attached at one end to the same fixed point, and they carry balls A and B, of masses $2m$ and m, at their other ends. A hangs vertically while B is held with its string at $60°$ to the vertical and released. Given that e for the two balls is $\frac{1}{2}$, show that A turns through an angle of arc $\cos\frac{7}{8}$ before coming to instantaneous rest after the first collision, and show that B turns through the same angle after the second collision.

24 A ball moving with a speed of u collides directly with an identical stationary ball. Given that $\frac{3}{8}$ of the original KE is lost at the collision, find the velocities just after the collision and the value of e.

25 A ball is dropped onto a horizontal plane and strikes it at 20 m s^{-1}. $e = \frac{3}{4}$. Find the time interval between the first collision and the moment at which the ball comes to permanent rest on the plane.

26 A ball is dropped from a height of 45 cm onto a horizontal plane. $e = \frac{1}{2}$. Prove that the ball travels a total distance of 75 cm before coming to permanent rest on the plane.

9

Rigid Bodies in Equilibrium. Friction

We have already considered in previous chapters some special cases of bodies in equilibrium. Chapter 4 dealt with bodies of negligible extension, or particles, while in Chapter 6 we investigated the equilibrium of bodies under systems of parallel forces. In this chapter we consider rigid bodies in equilibrium more generally, and since friction is often involved in such cases, this topic is also dealt with here. Before turning to friction, however, we must examine one more special case, namely that of bodies in equilibrium under *three* forces.

Theorem Three forces in equilibrium are either parallel or concurrent.

Proof If the three forces are not parallel let two of them meet at a point P. These two forces then have no moment about P; hence if the third force does not also pass through P there will be a resultant moment about P and the system cannot be in equilibrium. It follows that either all three forces pass through P, and are thus concurrent, or all three are parallel.

It follows from this theorem that a body in equilibrium under three non-parallel forces constitutes a special case. If a system of forces is concurrent, it is a system of the kind dealt with in Chapter 4, even though in that chapter the forces were always considered to act upon particles whereas we shall now be mainly concerned with extended bodies. This means that the triangle of forces law is applicable to this kind of case, and thus that Lami's theorem can be used. However the most useful application of the theorem just proved is that it often gives us the *direction* of a force whose direction would otherwise be unknown. This is illustrated by the first of the worked examples which now follow.

Worked examples – three force problems

Example 1 A uniform rod AB of mass 10 kg is smoothly hinged to a fixed support at A and held at an acute angle to the upward vertical by a string connecting B to a point C directly above A. If $AB = AC = 5$ units, and $BC = 6$ units, find (a) the tension in the string, (b) the magnitude and direction of the reaction at the hinge.

The direction of the force which a hinge applies to a body (the *reaction* of the hinge) is not in general known. Consequently the usual way to deal with a force of this kind is to express it in terms of horizontal and vertical components, denoted by X and Y. In this problem, however, we do know the direction of the reaction, since the above theorem tells us that it must pass through the point of intersection of the other two forces, namely the mid-point of *BC*. The force diagram is thus as shown in Fig. 9.1.

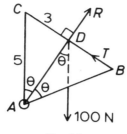

Fig. 9.1

(a) We can find *T* either by taking moments about *A* or resolving along *BC*, in both cases obtaining an equation in which *R* does not appear. The latter method is slightly quicker here.

Resolving along *BC*,

$$T = 100 \sin \theta = 100 \times \tfrac{3}{5} = \textbf{60 N}.$$

(b) Resolving along *AD*,

$$R = 100 \cos \theta = 100 \times \tfrac{4}{5} = \textbf{80 N}.$$

The reaction is clearly directed at an angle of arc sin $\tfrac{3}{5}$ to the upward vertical.

It should be realised that it does not always help in three-force problems to use the fact that the forces are concurrent. The next example illustrates this.

Example 2   A uniform rod *AB* of mass 10 kg is smoothly hinged to a fixed support at *A* and held at 60° to the upward vertical by means of a string attached at *B* which is perpendicular to the rod (Fig. 9.2). Find (a) the tension in the string, (b) the magnitude and direction of the reaction at the hinge.

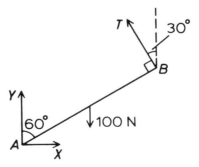

Fig. 9.2

Although we know that the reaction passes through the point of intersection of the forces of *T* and 100, this does not in itself (without some trigonometric calculations) give us the inclination of the reaction to the vertical. It is therefore better to represent the reaction as a pair of components.

(a) Letting the length of the rod be $2a$ and taking moments about A, we have

$$2aT = 100a \cos 30°,$$

from which

$$T = \textbf{43.30 N.}$$

(b) Knowing T, we can now find X and Y by resolving horizontally and vertically.

Resolving horizontally: $X = T \cos 60°$
$$= 21.65.$$

Resolving vertically: $Y + T \cos 30° = 100,$

from which

$$Y = 62.5.$$

Finally we find the magnitude and direction of the reaction by drawing a parallelogram of forces (Fig. 9.3).

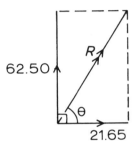

Fig. 9.3

By Pythagoras,

$$R^2 = 21.65^2 + 62.5^2$$
$$\therefore \ R = \textbf{66.14 N.}$$

Also

$$\tan \theta = \frac{62.50}{21.65}$$
$$\therefore \ \theta = \textbf{70.89°.}$$

This is the angle between the reaction and the horizontal.

In some three-force problems we are asked to derive or prove purely *geometrical* properties of the object or objects concerned. In such cases it is often possible to dispense with the usual techniques of resolving, taking moments, etc., and simply apply geometric and triogonometric methods to the figure we obtain by representing the three forces as concurrent. The final example illustrates this approach.

Example 3   A uniform rod ACB rests at $60°$ to the vertical with C in contact with a smooth horizontal rail. It is held in equilibrium by a horizontal force which acts at its lower end A. Prove that $AC:CB = 3:5$.

The three forces on the rod are its own weight, the reaction of the rail and the horizontal force at A. Since these are concurrent the diagram is as shown in Fig. 9.4.

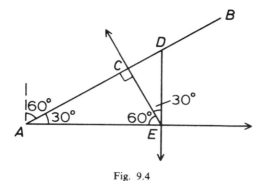

Fig. 9.4

Now from triangle ACE we have

$$\tan 60° = AC/CE$$
$$\therefore \; AC = CE\sqrt{3} \qquad \text{(using } \tan 60° = \sqrt{3}\text{).}$$

Also from triangle CDE we have

$$\tan 30° = CD/CE$$
$$\therefore \; CD = CE/\sqrt{3} \qquad \text{(using } \tan 30° = 1/\sqrt{3}\text{).}$$

It follows that

$$AC:CD = \sqrt{3}:1/\sqrt{3}$$
$$= 3:1.$$

We can therefore let $AC = 3$ units and $CD = 1$ unit, and then since D is the midpoint of the rod, $DB = 4$ units. Hence $AC:CB = 3:5$.

Exercise 9a

1 A uniform rod AB of mass 8 kg can turn in a vertical plane about a smooth hinge at its lower end A, and a string connects the other end B to a point C which is vertically above A. Given that $\angle BAC = 40°$ and $AB = AC$, find the tension in the string, the magnitude of the reaction of the hinge, and the angle the reaction makes with the upward vertical.

2 A uniform sphere of mass 600 g and radius 5 cm is held in contact with a smooth vertical wall by a string of length 8 cm whose ends are attached to a point on the sphere and a point on the wall. Find the tension in the string and the reaction of the wall.

3 A uniform rod of mass 3 kg is smoothly hinged to a support at its lower end and held at an acute angle to the upward vertical by a force of 10 N, at right angles to the rod, at the other. Find the angle between the rod and the vertical, and the magnitude and direction of the reaction of the hinge.

4 A uniform sphere of weight 20 N rests on a smooth plane of inclination 25°, being held in position by a horizontal string attached to a point on its surface. Find the tension in the string and the reaction of the plane.

5   A uniform rod of length 1 m and mass 2 kg is smoothly hinged at its lower end to a fixed support which is 60 cm from a smooth vertical wall. The upper end rests against the wall. Find the magnitudes of the reactions of the hinge and the wall.

6   A uniform rod of weight W is smoothly hinged to a fixed support at its lower end, and held at a certain angle to the vertical by a force of $0.4W$ which is perpendicular to the rod. If the reaction of the hinge acts along the rod, find the magnitude of this reaction and the angle between the rod and the vertical.

7   A uniform rod of length 60 cm and mass 500 g is suspended from a fixed point by means of two strings, of lengths 36 cm and 48 cm, which are attached to its ends. Find the tensions in the strings.

8   A uniform square lamina $ABCD$ of weight 10 N can turn in a vertical plane about a fixed support at A. Owing to a force of F acting at right angles to BC, it rests with B as its highest point, the angle between BA and the upward vertical being $60°$. Find F and the magnitude of the reaction of the support when (a) this reaction acts along AC, (b) F acts through C.

9   A uniform rod rests with its ends on two smooth perpendicular planes whose inclinations to the horizontal are θ and $90° - \theta$. Prove that one of the angles between the rod and the vertical is 2θ.

10   A uniform rod AB of length 40 cm and mass 50 g is smoothly hinged to a fixed support at its lower end A, and rests to $60°$ to the vertical over a smooth horizontal rail which is 10 cm from B. Find the reaction of the rail and the magnitude and direction of the reaction of the hinge.

11 $ABCD$ is a uniform rectangular lamina of mass 800 g in which $AB = 50$ cm and $BC = 20$ cm. The lamina can turn in a vertical plane about a smooth hinge at its lowest point A, and it rests with AB at $35°$ to the horizontal owing to a horizontal string attached at C. Find the tension in the string and the magnitude and direction of the reaction at A.

12   A uniform rod of mass 500 g is smoothly hinged to a fixed support at its lower end and held at $40°$ to the vertical by a force at its upper end. If the reaction of the hinge is horizontal, find the magnitude and direction of the force at the upper end.

13   A uniform rod of length $2a$ rests in equilibrium over a smooth horizontal rail owing to a horizontal string attached to its upper end. Find the distance of the rail from the upper end when the inclination of the rod to the horizontal is (a) $45°$, (b) $60°$.

14   A non-uniform rod AB of length 80 cm is smoothly hinged to a fixed support at its lower end A, and rests at $25°$ to the horizontal owing to a string attached at B which is inclined at $50°$ to the horizontal. Find to the nearest cm the distance of the centre of gravity of the rod from A if the reaction at A is horizontal.

15   A uniform rod of length 1 m and mass 120 g rests over a smooth horizontal rail with its lower end on a smooth plane. The inclinations to the horizontal of the rod and the plane are respectively $40°$ and $30°$. Find the reactions of the rail and the plane, and the distance, to the nearest cm, of the rail from the lower end of the rod.

16   A uniform rod rests at $30°$ to the horizontal with one end in contact with the inside surface of a smooth hemispherical bowl of radius a which is held with its rim horizontal. The rod rests against the rim and partially protrudes from the bowl. Find the length of the rod.

Friction

Whenever two solid bodies are in contact each exerts a force on the other at the surface of contact. The two forces are equal and opposite, and either can be resolved into a component perpendicular to the common surface and a component along this surface. The former, usually called the *normal reaction* and denoted by R, is equal to whatever external force is pressing the surfaces together, and it is present whenever there is genuine contact between bodies. The latter, if it exists at all, must be due to friction; it can only come into play in response to an attempt to slide one surface over the other, and it is always opposite in direction to the motion or attempted motion. We shall denote such a frictional force by F.

Now consider Fig. 9.5, which shows an attempt being made by means of an applied force P to slide a movable body along the surface of a fixed one. Suppose the force P starts at zero and is gradually increased. The following stages occur.

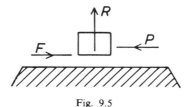

Fig. 9.5

1. The frictional force F comes into action and exactly balances P. No motion occurs.
2. F reaches the greatest value the two surfaces are capable of providing, given the particular value of R which occurs. This value of F is called *limiting friction*, and denoted by F_L. At this point the body is still not moving, but it is on the verge of moving, and a slight increase in F, or a tapping of the surfaces, will cause motion to occur. The body is said to be in *limiting equilibrium*.
3. The equilibrium is disturbed in one of the ways just described, and motion takes place. F now drops slightly to a value below F_L, called *sliding friction* or *dynamic friction*.

In this chapter we shall be frequently concerned with bodies in limiting equilibrium, and thus with cases in which the frictional force is F_L. Experiment in fact shows that for a particular pair of surfaces F_L is directly proportional to the normal force, so that we can write

$$F_L = \mu R.$$

It is clear from this equation that the greater the value of the constant of proportionality, μ, the greater is the value of F_L for a given value of R. Thus μ measures the friction-producing power of the pair of surfaces, and it is known as their *coefficient of friction*.

Angle of friction

Suppose we construct a parallelogram of forces to obtain the single resultant force which a fixed body exerts on a movable one when the latter is in limiting equilibrium (Fig. 9.6).

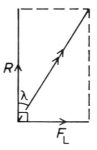

Fig. 9.6

The angle which the resultant makes with the normal to the common surface is called the *angle of friction* of the pair of surfaces, and denoted by λ. It is actually just an alternative to μ as a measurement of the friction-producing power of the surfaces; for the greater the value of μ, the greater is that of λ. The exact relationship can be obtained from Fig. 9.6. Since

$$\tan \lambda = F_L/R = \mu R/R,$$

we have

$$\boxed{\tan \lambda = \mu.}$$

We consider next some practical examples in which bodies are in equilibrium under the influence of frictional forces. The first result obtained – concerning a body on a rough inclined plane – is sufficiently important to be regarded as a theorem. It is proved for the case of a particle, but in fact holds for any body which is small enough to slide rather than topple.

Theorem   A particle on a rough plane of inclination θ is on the verge of sliding when $\tan \theta = \mu$.

Proof   When the particle is on the verge of sliding the friction is limiting; hence the forces on it are as shown in Fig. 9.7.

Now resolving at right angles to the plane,

$$R = W \cos \theta, \qquad (1)$$

and resolving along the plane,

$$R = W \sin \theta. \qquad (2)$$

Dividing (2) by (1), we obtain the required result, i.e.

$$\tan \theta = \mu.$$

(Note that since $\tan \lambda = \mu$, it also follows that the particle slips when the angle of inclination is equal to the angle of friction.)

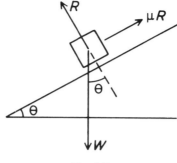

Fig. 9.7

Worked examples

Example 1 A 2 kg particle lies on a plane of inclination 60°. The coefficient of friction for particle and plane is 0.25. Find the least horizontal force which prevents the particle from sliding down the plane.

When the particle is on the verge of moving down the plane, the frictional force acts up the plane. Since the friction is limiting, this force can be expressed immediately as μR, i.e. in this case $0.25R$, (Fig. 9.8).

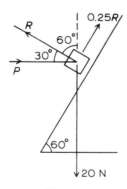

Fig. 9.8

There are two unknowns, namely P and R. We can obtain an equation which does not involve P, and from which we can therefore obtain R, by resolving vertically:

$$R \cos 60° + 0.25R \cos 30° = 20$$

that is,

$$R(0.5 + 0.2165) = 20,$$

from which

$$R = 27.91 \text{ N}.$$

Now, knowing R, we can find the required force P by resolving horizontally:

$$P + 0.25R \cos 60° = R \cos 30°$$
$$\therefore \ P = R(\cos 30° - 0.25 \cos 60°)$$
$$= \mathbf{20.68 \ N}.$$

(Note that if we required the least horizontal force which moves the particle up the plane, we should simply reverse the frictional force, then proceed in the same way.)

Example 2   A uniform ladder of weight W rests at an inclination to the horizontal of θ, where $\tan\theta = \frac{5}{4}$, with its foot on rough horizontal ground and its upper end against a rough wall. μ for both ground and wall is $\frac{1}{2}$. How far can a man of weight W climb up the ladder before it slips?

Let the length of the ladder be $2a$, and let the man be a distance x from its foot when it is on the point of slipping. Each frictional force can then be expressed as μ times the corresponding normal reaction (Fig. 9.9).

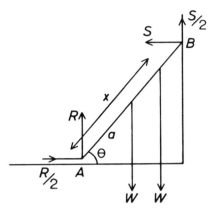

Fig. 9.9

Resolving horizontally: $R/2 = S$.
Resolving vertically: $R + S/2 = 2W$.
Eliminating R: $2S + S/2 = 2W$,

from which $S = 4W/5$.

Having decided to eliminate R, rather than S, it is appropriate to take moments about A rather than B, to obtain another equation which does not contain R.
Taking moments about A:

$$\frac{S}{2} \times 2a\cos\theta + S \times 2a\sin\theta = Wa\cos\theta + Wx\cos\theta.$$

Substituting $4W/5$ for S and dividing through by $\cos\theta$:

$$\frac{4Wa}{5} + \frac{8Wa\tan\theta}{5} = Wa + Wx.$$

Dividing through by W and substituting $\frac{5}{4}$ for $\tan\theta$:

$$\tfrac{4}{5}a + 2a = a + x,$$

from which $x = 9a/5$.
The man of weight W can therefore climb $\frac{9}{10}$ **of the length of the ladder** before it slips.

Example 3 A uniform ladder AB of weight W rests at an inclinaion of θ with A on rough ground and B in contact with a smooth wall. Express both the normal and frictional components of the force on the ladder at A in terms of W and θ, and hence show that the coefficient of friction is at least $\frac{1}{2}\cot\theta$.

We are not told that the friction at the ground is limiting, so we must express the frictional force by F and not μR. Since the wall is smooth it exerts only a normal force on the ladder, (Fig. 9.10).

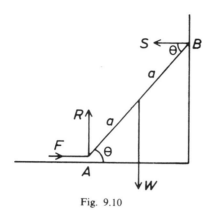

Fig. 9.10

We can express F in terms of W and θ by resolving horizontally and taking moments about A, thus obtaining two equations in which R does not occur.

$$F = S$$
$$Wa\cos\theta = S \times 2a\sin\theta.$$

Eliminating S from these two equations, we obtain

$$F = \tfrac{1}{2}W\cot\theta.$$

Since $R = W$ (resolving vertically), the normal and frictional forces are W and $\frac{1}{2}W\cot\theta$.

We obtain the least possible value of μ by supposing that the ladder is already on the point of sliding, so that $\mu = F/R$. The minimum possible value is thus $\frac{1}{2}W\cot\theta/W = \frac{1}{2}\cot\theta$.

A final example illustrates the usefulness of the idea of *angle of friction*.

Example 4 A sphere is held on an inclined plane by a force applied to the highest point of the sphere. Show that the angle of friction between sphere and plane is at least half the angle of inclination of the plane.

In all force diagrams involving limiting friction we have a choice between putting in a pair of components R and μR, and showing the resultant of these two forces together with the angle of friction. In the present case we obtain a three-force problem by adopting the latter procedure (Fig. 9.11).

Figure 9.11 represents the sphere as on the verge of slipping, so that the angle between the normal to the plane and the resultant force applied by the plane is λ. This value of λ is clearly the least value consistent with equilibrium; if λ is greater,

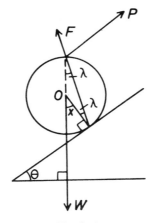

Fig. 9.11

the sphere can be held in position without being on the verge of sliding. P is the force applied to the top of the sphere; in this question its direction is not relevant.

Clearly $x = 2\lambda$, since O is the centre of the sphere. Also $x = \theta$, from the geometry of the figure. Hence $\lambda = \theta/2$, and thus the least possible value of the angle of friction is half the angle of inclination of the plane.

Toppling problems

The equilibrium of a body standing on a plane may be broken by applying a force to it or by tilting the plane. The manner in which the equilibrium is destroyed when this is done is not always apparent without calculation; usually the body can either slide along the plane or topple about one of its edges, and which in fact happens depends upon the position of the centre of gravity and the friction between the body and the plane. We give two examples to illustrate the kind of calculation involved: one very simple, one more complicated.

Example 1   A uniform cylinder whose height is $1\frac{1}{2}$ times its radius stands with one of its end faces in contact with a horizontal plane. The plane is steadily tilted. Show that if the cylinder topples before it slides, then μ for the cylinder and the plane is at least $\frac{4}{3}$.

The cylinder is on the point of toppling when its centre of gravity is directly above its lowest point, as shown in Fig. 9.12. At this point we clearly have

$$\tan \theta = r \div \frac{3r}{4} = \frac{4}{3}.$$

We know from the theorem proved above, however, that the cylinder slides when $\tan \theta = \mu$. Hence μ must be at least $\frac{4}{3}$ or the cylinder would have slipped before reaching the toppling position.

Example 2   A uniform rectangular block of weight W lies on a rough inclined plane as shown in Fig. 9.13. $AB = 2AD$, $\theta = \text{arc tan} \frac{3}{4}$, and $\mu = 2$. When a gradually increasing horizontal force of P is applied in the plane of the diagram at D, as shown, determine whether equilibrium is broken by the block's turning about the edge through A, or sliding down the plane.

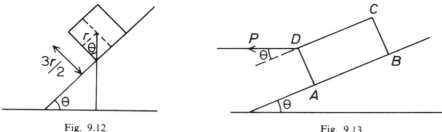

Fig. 9.12   Fig. 9.13

In this case the force diagram is best drawn with the forces of P and W already resolved into components along and perpendicular to the plane. The position of the resultant reaction, R, varies with P; when the block is on the verge of tipping it must act through A, as shown in Fig. 9.14. It is convenient to let $AB = 4$ and $AD = 2$.

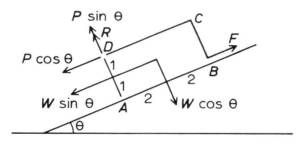

Fig. 9.14

The simplest method is to calculate both (a) the value of P at which the block slides, assuming that it does not topple, and (b) the value at which it topples, assuming that it does not slide. The lower of these two values is then the one which in fact breaks the equilibrium.

(a) Let the block be on the verge of sliding, so that $F = 2R$.

Resolving along the plane: $P \cos \theta + W \sin \theta = 2R$; that is,

$$\frac{4P}{5} + \frac{3W}{5} = 2R$$

or

$$4P + 3W = 10R. \qquad (1)$$

Resolving normally to the plane,

$$P \sin \theta + R = W \cos \theta$$

that is,

$$\frac{3P}{5} + R = \frac{4W}{5}$$

or
$$3P + 5R = 4W. \qquad (2)$$

Eliminating R from (1) and (2),
$$4P + 3W = 2(4W - 3P);$$
that is,
$$4P + 3W = 8W - 6P,$$
$$\therefore \ P = W/2.$$

(b) Let the block be on the point of toppling

Taking moments about A,
$$2W \cos \theta = W \sin \theta + 2P \cos \theta;$$
that is,
$$\frac{8W}{5} = \frac{3W}{5} + \frac{8P}{5}$$
$$\therefore \ 5W = 8P$$
$$\therefore \ P = 5W/8.$$

Since $W/2$ is less than $5W/8$ it follows that the block slides rather than topples.

We conclude this chapter by giving just one example of a somewhat more complicated nature, involving a pair of jointed rods.

Example   Two uniform rods, AB and BC, of equal length but with weights of W and $3W$, respectively, are smoothly hinged at B and rest in a vertical plane with their lower ends A and C in contact with rough horizontal ground. Given that $\angle BAC = 60°$, find in terms of W the normal and frictional forces at A and C, and the magnitude of the reaction at the hinge. Find also the minimum possible value of the coefficient of friction, μ, between the rods and the ground.

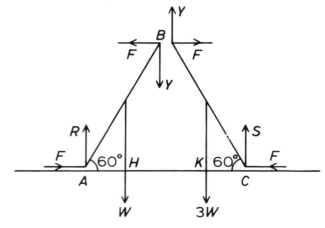

Fig. 9.15

This question involves *two* movable objects, and we need to show the forces on each. We therefore draw them slightly apart (Fig. 9.15). With regard to this diagram note the following:

(a) The two forces of F at B are equal and opposite by Newton's third law, as are the two forces of Y.
(b) The horizontal forces at A and C are also equal to F, since there is no net horizontal force on either rod.
(c) The two forces of Y may not be correct in direction; if this is so, Y simply turns out to have a negative value.

We can find R and S by considering *the whole system*. The forces at the hinge then become internal and can be ignored.

Let $AH = KC = 1$ unit of length; then $HK = 2$ units. Taking moments about A for the whole system,

$$4S = 9W + W$$
$$\therefore\ S = 5W/2.$$

Resolving vertically for the whole system,

$$R + S = 4W$$
$$\therefore\ R = 3W/2.$$

Now taking moments about B for rod AB,

$$2F \sin 60° + W \cos 60° = 2R \cos 60°;$$

that is,

$$F\sqrt{3} + W/2 = R,$$

and since $R = 3W/2$, $F = W/\sqrt{3}$ or $F = W\sqrt{3}/3$.

In order to find the reaction at the hinge we require Y. Hence, resolving vertically for rod AB:

$$Y + W = R$$
$$\therefore\ Y = W/2.$$

Letting the magnitude of the reaction at the hinge be P, we thus have

$$P^2 = F^2 + Y^2$$
$$= \frac{W^2}{3} + \frac{W^2}{4} = \frac{7W^2}{12}.$$

Hence

$$P = 0.7638\,W.$$

To find the minimum possible value of μ, we first calculate F/R and F/S:

$$\frac{F}{R} = \frac{W\sqrt{3}}{3} \times \frac{2}{3W} = \frac{2\sqrt{3}}{9}$$

$$\frac{F}{S} = \frac{W\sqrt{3}}{3} \times \frac{2}{5W} = \frac{2\sqrt{3}}{15}.$$

Now rod AB is on the point of sliding when $F/R = \mu$, and rod BC is on the point of sliding when $F/S = \mu$. Since neither rod in fact slides, μ must have at least the *greater* of these two values, namely $2\sqrt{3}/9$.

Exercise 9b

Points to note

(a)   It is essential in problems involving bodies in equilibrium to draw large, clear force diagrams.

(b)   Frictional forces can be expressed in the form μR only if the friction is limiting.

In the examples which follow, coefficients of friction will be denoted by μ.

1   A desk-lid is steadily raised, and when it is at $25°$ to the horizontal a book begins to slide. Find μ for book and lid.

2   A 5 kg particle lies on a plane of inclination $40°$. Given that $\mu = \frac{1}{2}$, find the least force acting straight up the plane which (a) prevents the particle from sliding down, (b) moves the particle up the plane.

3   A 2 kg particle lies on a horizontal plane. Find μ if a force of 25 N acting away from the plane at $30°$ to the horizontal just causes the particle to move.

4   A 5 kg body rests on a plank which is 2 m long and fixed with its upper end 120 cm above its lower end. $\mu = \frac{1}{4}$. Find the least horizontal force which will move the body up the plank.

5   A horizontal force of 80 N is just sufficient to move a 3 kg body up a plane of inclination $50°$. Find μ.

6   A 20 kg body lies on horizontal ground. The angle of friction is $60°$. Find the least force acting at $30°$ to the ground which will move the body.

7   Two surfaces for which the angle of friction is $50°$ are pressed together. Find the force doing this if it requires a force of 25 N to slide the surfaces over each other.

8   A uniform cube rests on a desk-lid, with one of its edges parallel to the hinges. Find the minimum possible value of μ if the cube topples before it slides when the lid is raised.

9   A 4 kg body lies on a plane of inclination $20°$. If a force of 30 N acting away from the plane at $50°$ to the horizontal will just move the body up the plane, what is the value of μ?

10   A uniform cylinder whose height is equal to its radius stands with an end-face in contact with a horizontal surface. The surface is steadily tilted, and it is found that the cylinder topples before it slides. Find the minimum possible value of μ.

11   A uniform ladder of mass 20 kg leans in limiting equilibrium at $40°$ to the horizontal with its foot on rough horizontal ground and its upper end against a smooth vertical wall. Find μ for the ground and the ladder.

12   A uniform cylinder of radius r stands with an end-face in contact with a desk-lid which is steadily raised. $\mu = \frac{2}{3}$. Given that the cylinder slides before it topples, find its maximum possible height.

13   A uniform ladder rests with its lower end on rough horizontal ground and its upper end against a smooth vertical wall. μ for the ladder and the ground is 0.3. Find the greatest possible angle of inclination of the ladder to the wall.

14   A uniform rod AB of mass 4 kg leans at $50°$ to the horizontal with A on horizontal ground and B in contact with a vertical wall. $\mu = \frac{1}{2}$ for both the ground and the wall. Find the downward force at B which is just sufficient to make the rod slip.

15   A uniform ladder AB of weight W leans at $60°$ to the horizontal with A on rough horizontal ground and B against a smooth vertical wall. μ for the ladder and the ground is 0.4. How far can a man of weight $2W$ climb up the ladder before it slips?

16 A uniform ladder AB of weight W leans at θ to the horizontal with A on rough horizontal ground and B against a smooth vertical wall. A man of weight W stands $\frac{1}{4}$ of the way up the ladder. Obtain expressions in terms of W and θ for the normal and frictional components of the force of the ground on the ladder. Deduce the minimum possible value of μ.

17 A uniform rod AB of weight W leans at θ to the horizontal with its foot A on rough horizontal ground and B resting on a smooth horizontal rail. Obtain expressions in terms of W and θ for the normal and frictional components of the force of the ground on the rod, and deduce the minimum possible value of μ.

18 A uniform rod rests with its lower end on rough horizontal ground and its upper end against a smooth vertical wall. λ is the angle of friction between the rod and the ground and θ is the angle between the rod and the vertical. Prove *geometrically* (using the concurrence of three forces theorem) that when the rod is on the point of slipping, $\tan\theta = 2\tan\lambda$.

19 A uniform rod ACB in which $AC = \frac{1}{4}AB$ rests at an angle of arc $\tan\frac{3}{4}$ to the vertical with its lower end A in contact with a rough wall. At C the rod is supported by a smooth peg. Find the minimum possible value of μ for the rod and the wall.

20 A uniform rod of length 40 cm and weight 20 N rests in limiting equilibrium with its lower end in contact with a rough plane of inclination $30°$, μ for the rod and the plane being 2. The rod is in the same vertical plane as the line of greatest slope through its lowest point and is inclined at $30°$ to the upward direction of this line, being held in position by a string which is perpendicular to the rod. Find the tension in the string and the distance between the point of attachment of the string and the lower end of the rod.

21 A uniform rectangular block stands on a horizontal surface and a steadily increasing horizontal force is applied to the top, in the direction of an edge of length l. The height of the block is h. Find the condition in terms of μ, l and h that the block topples rather than slides.

22 A uniform ladder of length $4a$ and weight W leans at $60°$ to the horizontal with its foot on rough horizontal ground and its upper end against a smooth vertical wall. A man of weight $2W$ is $\frac{3}{4}$ of the way up the ladder and μ for the ladder and ground is $\sqrt{3}$. What couple would be just sufficient to make the ladder slip downwards?

23 A uniform rod rests at $30°$ to the horizontal with its lower and upper ends in contact with rough horizontal ground and a smooth plane of inclination $60°$. Find the minimum possible value of μ for the rod and the ground.

24 A uniform cube of weight W stands on a plane of inclination $20°$ with all its edges parallel or perpendicular to a line of greatest slope. $\mu = 0.3$. A gradually increasing horizontal force, tending to pull the cube up the plane, is applied to the highest edge. Determine whether the cube slides or topples, and find the value of the force which achieves this.

25 A uniform ladder of length $4a$ and weight W rests in limiting equilibrium at an angle of arc $\tan\frac{4}{3}$ to the horizontal with its ends in contact with horizontal ground and a vertical wall which are equally rough. A man of weight $2W$ stands $\frac{1}{4}$ of the way up. Find the value of μ.

26 Two step ladders AB and BC, each of length $4a$ and weight W, are smoothly hinged at B and stand with their ends A and C on rough horizontal ground. $\angle BAC = \text{arc tan}\frac{4}{3}$. When a man of weight W stands on AB at a distance a from B, determine the normal and frictional forces at A and C, and the minimum possible value of μ.

27   Two uniform rods AB and BC, each of weight W but with lengths in the ratio $4:3$, are smoothly jointed at B and rest at right angles to each other in a vertical plane with their lower ends A and C in contact with rough horizontal ground. Find the minimum possible value of μ for the rods and the ground.

28   Two uniform rods AB and BC, of equal lengths $2a$ but with weights W and $2W$, are smoothly jointed at B. The ends A and C are smoothly jointed to fixed points on the same horizontal level, and the system hangs with B a distance a below this level. Find the magnitude and direction of the reaction at B.

10

Motion of Connected Bodies

The principles involved in this topic have all been explained in previous chapters, and we shall now deal with it purely by means of worked examples. The topic has been left until this stage because problems on connected bodies often incidentally involve friction, resolution and impulsive forces, as the examples will show.

   All pulleys in the examples will be regarded as light and frictionless. No force will then be required to accelerate such a pulley, and the tensions in the sections of string on each side of the pulley will therefore be equal. Also the strings will all be considered light and inextensible, and all surfaces may be taken to be frictionless unless otherwise stated.

Worked examples

Example 1   In the system shown in Fig. 10.1 find the acceleration, the tension in the string, and the force on the pulley due to the string.

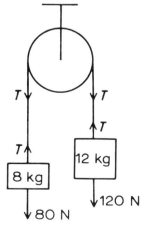

Fig. 10.1

First method

In this method we begin by applying the equation $F = ma$ to the whole system, and thereby obtain the acceleration. Then, knowing the acceleration, we find the tension by applying $F = ma$ to either of the bodies alone.

Consider the whole system. We have

$$\text{net external accelerating force} = 120 - 80 = 40 \text{ N},$$
$$\text{total mass} = 20 \text{ kg}.$$

Hence, from $F = ma$, we have $a = 40/20 = 2 \text{ m s}^{-2}$.
Now consider the 8 kg body alone. Applying $F = ma$ just to this body and remembering that it accelerates upwards, we have

$$T - 80 = 8 \times 2$$
$$\therefore T = 96 \text{ N}.$$

Clearly the force the string applies to the pulley is simply $2T$, that is, **192 N**.

Second method

This consists in applying $F = ma$ to each body separately, thus obtaining two simultaneous equations in a and T.

$$\text{Taking the } 8 \text{ kg body:} \quad T - 80 = 8a \quad (1)$$
$$\text{Taking the } 12 \text{ kg body:} \quad 120 - T = 12a \quad (2)$$

$$(1) + (2): 40 = 20a$$
$$\therefore a = 2 \text{ m s}^{-2}.$$

Substituting this value into (1):

$$T - 80 = 16$$
$$\therefore T = 96 \text{ N}.$$

Example 2   Suppose that the whole system of Example 1 accelerates from rest for 0.6 s, and then the 8 kg body picks up a 10 kg body, previously at rest. Find the time from that moment to the point at which the system comes to instantaneous rest.

First we find the velocity v_1 of the system just before picking up the extra weight. Using

$$v = u + at,$$

we have

$$v_1 = 2 \times 0.6$$

that is

$$v_1 = 1.2 \text{ m s}^{-1}.$$

Now picking up the extra weight involves an impulsive force since both the weight and the system must undergo effectively instantaneous changes in velocity. We therefore use the principle of conservation of momentum to find the velocity v_2 of the system just after the weight is picked up.

$$20 \times 1.2 = 30v_2$$
$$\therefore v_2 = 0.8 \text{ m s}^{-1}.$$

Next we require the new acceleration of the system, which will be negative since the total mass on the left is now greater than that on the right. (The clockwise direction has been implicitly taken as positive since all the motion and acceleration has so far been in that direction.)

Net decelerating force on whole system $= 180 - 120$
$$= 60 \text{ N},$$
$$\text{total mass} = 30 \text{ kg}$$
$$\therefore \ a = -60/30 = -2 \text{ m s}^{-2}.$$

Finally we use $v = u + at$ to find the time the system takes to come to rest:
$$0 = 0.8 - 2t$$
$$\therefore \ t = 0.4 \text{ s}.$$

Example 3   In the system shown in Fig. 10.2 find the tensions in all the strings.

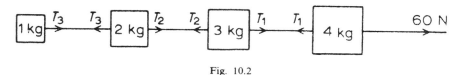

Fig. 10.2

It would be inappropriate here to use simultaneous equations, since there would be 4 of these. Consequently we begin by finding the acceleration.
   Consider the whole system. Using $F = ma$, we have
$$a = \frac{\text{external force}}{\text{total mass}} = \frac{60}{10} = 6 \text{ m s}^{-2}.$$

Now to obtain T_1 the best method is to apply $F = ma$ to the system consisting of the 1, 2 and 3 kg bodies combined. This gives immediately
$$T_1 = 6 \times 6 = \textbf{36 N}.$$

Similarly, to find T_2 we consider the 1 and 2 kg bodies together, and to find T_3 we consider the 1 kg body alone:
$$T_2 = 3 \times 6 = \textbf{18 N}$$
$$T_3 = 1 \times 6 = \textbf{6 N}.$$

Example 4   In the system shown in Fig. 10.3 find the acceleration, the tension, the force exerted by the string on the pulley, and the reaction between the pan and the weight.

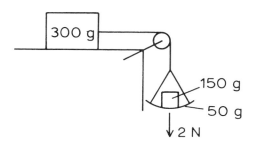

Fig. 10.3

In this system the weight of the 300 g body is at right angles to the direction of the motion, and since the surface is smooth this force does not influence that motion. The only external driving force is therefore the combined weight of the pan and the body it carries, i.e. 2 N. We thus have

$$external\ accelerating\ force = 2\ N,$$
$$total\ mass = 0.5\ kg$$
$$\therefore\ a = F/m = 2/0.5 = \mathbf{4\ m\,s^{-2}}.$$

We can now find the tension by applying $F = ma$ to the 300 g body alone:

$$T = 0.3 \times 4 = \mathbf{1.2\ N}.$$

The force on the pulley is the resultant of the two perpendicular forces of T which act on it. That is,

$$force\ on\ pulley = \sqrt{T^2 + T^2}$$
$$= T\sqrt{2}$$
$$= \mathbf{1.2\sqrt{2}\ N}.$$

To find the reaction between the pan and the 150 g weight we must either consider the pan alone or the weight alone. The latter procedure is slightly quicker since the pan has three forces on it whereas the weight has only the two shown in Fig. 10.4.

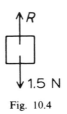

Fig. 10.4

Applying $F = ma$ to the weight, we have

$$1.5 - R = 0.15 \times 4,$$

from which

$$R = 0.9\ N.$$

Example 5   Suppose that in the system of Example 4 the pan is 50 cm above the ground and the 300 g weight is 1 m from the pulley. If the system starts from rest, how long will the 300 g body take to reach the pulley?

First we find the time it takes for the pan to reach the ground. Since

$$u = 0, \quad s = \tfrac{1}{2} \quad and \quad a = 4,$$

we have, using $s = ut + \tfrac{1}{2}at^2$,

$$\tfrac{1}{2} = \tfrac{1}{2} \times 4 \times t^2$$
$$\therefore\ t^2 = \tfrac{1}{4}$$
$$\therefore\ t = \tfrac{1}{2}\ s.$$

At this point the speed is given by

$$v = u + at$$
$$= 4 \times \tfrac{1}{2}$$
$$= 2 \text{ m s}^{-1}.$$

After the pan has come to rest on the ground the 300 g body has no forces on it in the direction of its motion, so it carries on moving with a constant speed of 2 m s^{-1}. The time it takes to cover the remaining $\tfrac{1}{2}$ m can therefore be obtained by applying the equation $s = ut$:

$$\text{time of second stage} = s/u = \tfrac{1}{2}/2 = \tfrac{1}{4} \text{ s}.$$

The total time is therefore $\tfrac{1}{2} + \tfrac{1}{4} = \tfrac{3}{4} \text{ s}$.

Example 6 Two rough inclined planes meet at right angles, the inclination of one to the horizontal being arc tan $\tfrac{3}{4}$ and that of the other arc tan $\tfrac{4}{3}$. Bodies of mass 2 kg and 4 kg lie on the respective planes, and the two are joined by a string passing over a pulley at the intersection of the planes. $\mu = \tfrac{1}{4}$ for both bodies. Find the acceleration of the system and the tension in the string.

Figure 10.5 shows all the forces relevant to the motion.

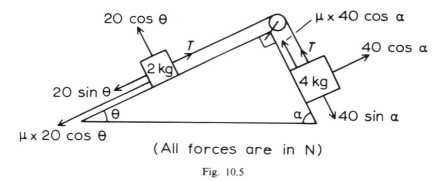

(All forces are in N)

Fig. 10.5

Note that since the motion is clearly clockwise both frictional forces are anticlockwise. Now since

$$\tan \theta = \cot \alpha = \tfrac{3}{4},$$

we have

$$\sin \theta = \cos \alpha = \tfrac{3}{5}$$

and

$$\cos \theta = \sin \alpha = \tfrac{4}{5}.$$

Hence net external accelerating force on whole system

$$= 40 \sin \alpha - \mu \times 40 \cos \alpha - 20 \sin \theta - \mu \times 20 \cos \theta$$
$$= 32 - 6 - 12 - 4$$
$$= 10 \text{ N}.$$

Total mass $= 6$ kg.

Hence, using $F = ma$, we have $a = 10/6 = 1\tfrac{2}{3} \text{ m s}^{-2}$.

Now to obtain T we apply $F = ma$ to one of the bodies alone. Taking the 2 kg body, we have

$$T - 20 \sin \theta - \mu \times 20 \cos \theta = 2 \times \tfrac{10}{6}$$

that is,

$$T - 12 - 4 = \tfrac{10}{3},$$

from which

$$T = 19\tfrac{1}{3} \text{ N}.$$

Exercise 10

In the following examples all pulleys and strings are weightless, all strings are inextensible, and all pulleys and surfaces are smooth unless otherwise stated.

1   Bodies with masses 2 kg, 3 kg, 5 kg, 10 kg are joined by strings and lie in a straight line on a horizontal table. A pull of 40 N, along this line, is applied to the 10 kg body. Find the acceleration and the tensions in all the strings.

2   Bodies of mass 30 kg and 10 kg are connected by a string passing over a pulley. Find the acceleration, the tension in the string, and the force on the pulley.

3   Bodies of mass 1 kg, 3 kg, 4 kg are connected by strings and lie in a straight line on a horizontal table. A pull of 60 N, along this line, is applied to the 1 kg body. Find the acceleration and the tensions in the strings.

4   Bodies of mass 5 kg and 3 kg are connected by a string passing over a pulley. Find the acceleration, the tension, and the force on the pulley.

5   A 12 kg body on a horizontal table is connected by a string passing over a pulley at the edge of the table to an 8 kg body which hangs vertically. Find the acceleration, the tension, and the force on the pulley.

6   Repeat question 5, interchanging the positions of the bodies.

7   Bodies A and B, with masses 3 kg and 2 kg, are connected by a straight string and lie on a rough horizontal surface. $\mu = 0.4$ for both bodies. A force of 30 N is applied to A in the direction BA. Find the acceleration and the tension.

8   A 500 g body on a horizontal table is connected by a string passing over a pulley at the edge of the table to a 200 g pan which carries a 300 g weight. Find the acceleration and the reaction between the pan and the weight.

9   Bodies of mass 6 kg and 4 kg are connected by a string passing over a pulley. They are released from rest and after $\tfrac{1}{2}$ s the 4 kg body picks up a stationary 5 kg body. Find the further distance travelled before the system comes to instantaneous rest.

10   A 8 kg body lying on a plane inclined at 30° to the horizontal is joined by a string passing over a pulley at the top of the plane to a 12 kg body which hangs vertically. Find the acceleration, the tension, and the force on the pulley.

11   Bodies A, B, C, with masses 2 kg, 3 kg, 5 kg, are joined by strings and lie in a straight line on a rough horizontal table. $\mu = 0.6$ for all bodies. Pulls of 120 N and 40 N are applied in opposite directions to A and C, respectively. Find the acceleration and the tensions in both strings.

12   A 30 kg body P, lying on a horizontal table, is connected by a taut string passing over a pulley at the edge of the table to a 10 kg body Q which hangs vertically. Q is 20 cm above a stop and P is 50 cm from the pulley. When the system is released from rest, find the total time it takes for P to reach the pulley.

13   Bodies of mass 2 kg and m kg are connected by a string passing over a pulley. Find m (a) if the body with this mass accelerates downwards at 4 m s^{-2}, (b) if the tension in the string is 16 N.

14   A 20 kg body on a rough horizontal table is connected by two strings passing over pulleys at opposite edges of the table to bodies with masses 10 kg and 50 kg which hang vertically. $\mu = 0.8$. Find the acceleration and the tensions in both strings.

15   A pan of mass 200 g carrying a 600 g weight is connected by a string passing over a pulley to a 400 g weight. Find the acceleration and the reaction between the pan and the weight.

16   Given that the system of question 15 starts from rest and the 600 g weight is removed from the pan after $\frac{1}{2}$ s, find the total time it takes for the system to return to its original position.

17   Bodies A and B, with masses 300 g and 200 g, are joined by a string 64 cm long. They are placed on a table 32 cm high, with B at an edge and the string straight and at right angles to the edge. If B is gently pushed over the edge find (a) the time it takes B to reach the ground, (b) the further time it takes A to reach the edge.

18   Bodies of mass m and $2m$ are connected by a string which passes over a pulley. Find the tension in the string in terms of m and g. What mass must be added to the smaller body to double this tension?

19   A 15 kg body on a horizontal table is connected by a taut string passing over a pulley at the edge of the table to a 1 kg pan which hangs vertically. A 4 kg body is dropped onto the pan from a height of 1.8 m, the system being released from rest just before the body hits the pan. Find the speed of the system $\frac{1}{2}$ s after the impact.

20   A body of mass $6\,m$ is connected by a string passing over a pulley to a pan of mass m carrying a weight of mass $3\,m$. Find the reaction between pan and weight in terms of m and g.

21   Two rough planes inclined at $30°$ and $60°$ to the horizontal meet along their top edges. Bodies with masses 10 kg and 40 kg stand on the respective planes and are joined by a string passing over a pulley at the top edge. $\mu = \frac{1}{2}$ for both bodies. Find the acceleration and the tension in the string.

22   Bodies A and B, with masses 30 kg and 20 kg, are joined by a string and lie with B above A on a line of greatest slope of a rough plane inclined at arc tan $\frac{4}{3}$ to the horizontal. $\mu = \frac{1}{3}$ for both bodies. Another string attached to B passes over a pulley at the top of the plane and supports at the other end a 25 kg body which hangs vertically. Determine the direction of motion of the 25 kg body and find its acceleration and the tensions in both strings.

23   A 500 g pan is connected by a string passing over a pulley to a 750 g weight. The system is released from rest. After $\frac{1}{2}$ s a weight of 750 g is dropped onto the pan, striking it at a speed of 3 m s^{-1}. Find the position of the pan relative to its starting point after another second.

11

Motion in a Circle

The relationship between linear speed and angular speed

Consider a particle moving in a circle of radius r with constant speed v, and suppose that in a time t it travels an actual or linear distance l while turning through θ radians (Fig. 11.1). We have $v = l/t$, and angular speed $\omega = \theta/t$. Since $l = r\theta$, therefore, the relationship between v and ω is

$$v = r\omega$$ (ω being in radians per unit time)

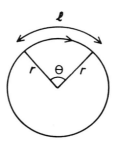

Fig. 11.1

Centripetal acceleration and centripetal force

According to Newton's first law, a body left to itself either remains at rest or moves in a straight line at constant speed. It follows that any body moving in a curve must have a resultant force acting on it, and thus be accelerating. We shall show that a particle moving at a constant speed of v in a circle of radius r has in fact an acceleration *towards the centre* of v^2/r. This is called a *centripetal* acceleration.

In Fig. 11.2, O is the centre and P and Q are two successive positions of the particle. We shall calculate the components of the acceleration at P in the tangential direction AB and the radial direction PO.

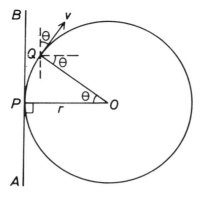

Fig. 11.2

Tangential component

At Q the component of velocity in direction AB is $v \cos \theta$, and at P the component in this direction is v. Hence

$$\text{increase in velocity} = v \cos \theta - v$$
$$= -v(1 - \cos \theta)$$
$$= -v \times 2 \sin^2 \tfrac{1}{2}\theta.$$

Also
$$\text{time taken} = PQ/v = r\theta/v.$$

Hence

$$\frac{\text{increase in velocity}}{\text{time taken}} = \frac{-2v^2 \sin^2 \tfrac{1}{2}\theta}{r} \frac{}{\theta} = -\frac{v^2}{r} \frac{\sin\tfrac{1}{2}\theta}{\tfrac{1}{2}\theta} \sin\tfrac{1}{2}\theta.$$

Now the tangential acceleration at P is the limit which this quantity approaches as θ tends to zero. Since

$$\lim_{x \to 0} \frac{\sin x}{x} = 1 \qquad \text{and} \qquad \lim_{x \to 0} \sin x = 0,$$

we have
$$\text{tangential component} = 0.$$

Radial component

At Q the component of velocity in direction PO is $v \sin \theta$, and at P the component in this direction is 0. Hence

$$\text{increase in velocity} = v \sin \theta,$$

and since, as before, the time taken is $r\theta/v$, we have

$$\frac{\text{increase in velocity}}{\text{time taken}} = \frac{v^2}{r} \frac{\sin \theta}{\theta}.$$

The limit of this quantity as θ tends to zero is v^2/r. This is the radial component of the acceleration at P, and as the tangential component is zero it is in fact the complete acceleration at P.

By using the relationship $v = r\omega$, we can also express the radial acceleration in terms of ω:

$$\frac{v^2}{r} = \frac{r^2\omega^2}{r} = r\omega^2.$$

Since the direction of the resultant acceleration at P is towards the centre, it can be described as *centripetal*. Hence, summing up, we have

$$\boxed{\text{centripetal acceleration} = \frac{v^2}{r} = r\omega^2.}$$

As explained above, any body moving in a curve must have a resultant force on it, even if it is moving at constant speed. In the case of a body moving in a circle at constant speed, there must be a resultant force on the body towards the centre to provide the centripetal acceleration. Such a force is called a *centripetal force*. For a body of mass m, the equation $F = ma$ gives

$$\boxed{\text{centripetal force} = \frac{mv^2}{r} = mr\omega^2.}$$

Some simple worked examples will now be given involving the motion with constant speed of a particle in a circle.

Example 1   A small body rests on a horizontal turntable at a distance of 20 cm from the centre (Fig. 11.3). Given that the body does not slide when the turntable is rotated at 33 revolutions per minute, find the minimum possible value of the coefficient of friction for the body and the turntable.

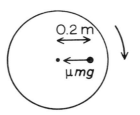

Fig. 11.3

First we convert the revs min^{-1} to rad s^{-1}:

$$33 \text{ revs min}^{-1} = 33 \times 2\pi \text{ rad min}^{-1}$$
$$= 33 \times 2\pi/60 \text{ rad s}^{-1}$$
$$= 3.456 \text{ rad s}^{-1}.$$

In this case the centripetal force is the frictional force. We obtain the minimum possible value of μ if we let the body be on the point of slipping, so that the frictional force can be represented as μR. Here R is simply the weight of the body, mg, and we therefore have

$$\text{centripetal force} = \mu mg$$

that is,

$$mr\omega^2 = \mu mg$$

that is,

$$0.2 \times 3.456^2 = 10\mu \qquad \text{(cancelling the } m\text{'s)}$$

from which $\mu = 0.2389$.
This is the required minimum possible value of μ.

Example 2   Particles A and B, with masses 5 kg and 3 kg, are connected by a light inextensible string which passes through a hole in a table. A hangs vertically, below the table, while B describes a circle of radius 40 cm on the surface of the table, which is smooth. Find the velocity of B if A is just held at rest (Fig. 11.4).

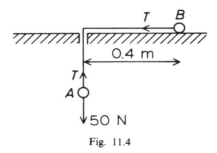

Fig. 11.4

Here the centripetal force is the tension in the string, which must be 50 N if A is just to be supported. Hence we have

$$\frac{mv^2}{r} = 50$$

that is,

$$\frac{3v^2}{0.4} = 50$$

from which $v = 2.582 \text{ m s}^{-1}$.

Example 3   A particle is attached to one end of an elastic string of natural length 40 cm. When the string is held vertically, the particle stretches it to a length of 60 cm. If the free end of the string is attached to a point on a smooth horizontal table, and the particle describes horizontal circles with an angular speed of 4 rad s^{-1}, find the extension of the string.

Here we begin by finding the modulus of elasticity λ of the string in terms of the mass m of the particle. When the string is vertical we have tension $T = mg = 10\,m$, extension $e = 0.2$, natural length $l = 0.4$. Hence, using

$$T = \frac{\lambda e}{l},$$

we have

$$10m = \frac{\lambda \times 0.2}{0.4},$$

for which

$$\lambda = 20m.$$

When the string is horizontal the tension is the centripetal force. Hence, letting the required extension be x, we have

$$\frac{\lambda x}{l} = mr\omega^2$$

that is,

$$\frac{20\,mx}{0.4} = m\,(0.4 + x)\,4^2$$

that is,

$$50x = 6.4 + 16x$$

from which

$$x = 0.1882\,\text{m} \quad \text{or} \quad \textbf{18.82 cm}.$$

The conical pendulum

Consider a light string which is attached at one end to a fixed point and at the other to a particle. If the particle moves in a horizontal circle at constant speed under the combined influence of its own weight and the tension in the string, we have the system known as the conical pendulum. Two worked examples are given to illustrate methods of dealing with this system.

Example 1   The string of a conical pendulum is 75 cm long and its tension is twice the weight of the particle (Fig. 11.5). Find the angular speed.

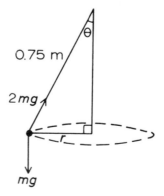

Fig. 11.5

Since the resultant acceleration is towards the centre of the horizontal circle and equal to $mr\omega^2$, we have, resolving horizontally,

$$2\,mg \sin \theta = mr\omega^2.$$

Dividing through by m and substituting $r/0.75$ for $\sin\theta$:

$$2\,g \times \frac{r}{0.75} = r\omega^2.$$

Hence
$$\omega^2 = 20/0.75,$$

from which
$$\omega = \mathbf{5.164\ rad\ s^{-1}}.$$

Example 2 The length of the string of a conical pendulum is 50 cm (Fig. 11.6). Find the speed of the particle if it rotates 30 cm below the string's upper end.

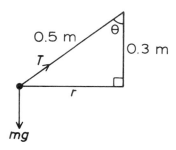

Fig. 11.6

We can obtain two equations from which both of the unknowns T and m can be eliminated by resolving horizontally and vertically:

$$T \sin\theta = \frac{mv^2}{r} \quad (1)$$

$$T \cos\theta = mg. \quad (2)$$

Dividing (1) by (2),

$$\tan\theta = \frac{v^2}{rg}.$$

Now clearly $r = 0.4$ m (3, 4, 5 triangle); hence $\tan\theta = 4/3$. We thus have

$$\frac{4}{3} = \frac{v^2}{0.4 \times 10},$$

from which
$$v = 2.309\ \text{m s}^{-1}.$$

Other cases of the horizontal motion of a particle in a circle

An example of horizontal motion which is identical in all essential respects to the conical pendulum is that of a particle moving at constant speed round the inside of a smooth hemispherical bowl. The normal reaction R takes the place of the tension in the string, and the radius of the bowl, a, corresponds to the length of the string (Fig. 11.7).

Somewhat more complicated are cases in which a particle is attached to two strings, or a smooth ring is threaded onto a single string. The final worked example is of this kind.

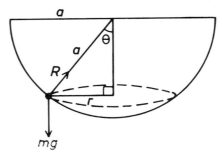

Fig. 11.7

Example 3   Strings AB, BC, of lengths $4a$, $3a$, are attached to a particle at B, and the ends A, C are held fixed with A a distance $5a$ directly above C. Find (a) the minimum speed at which B can rotate with both strings straight, (b) the tension in the lower string when the speed has twice this value.

(a) The minimum speed occurs when the lower string is straight but slack, thus having a tension of zero. We therefore have in effect a simple conical pendulum (Fig. 11.8).

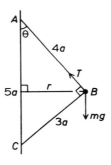

Fig. 11.8

As in Example 2 above, we have

$$\tan \theta = \frac{v^2}{rg}$$

that is,

$$\frac{3}{4} = \frac{v^2}{4ag \sin \theta}$$

from which

$$v^2 = \frac{9\,ag}{5}.$$

The minimum speed is thus $3\sqrt{\dfrac{ag}{5}}$.

(b) v is now $6\sqrt{ag/5}$ so that $v^2 = 36\,ag/5$. As before, $r = 4a \sin \theta = 12a/5$. We require T_2 (Fig. 11.9).

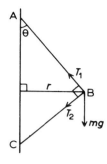

Fig. 11.9

Resolving horizontally,

$$T_1 \sin \theta + T_2 \cos \theta = \frac{mv^2}{r} = \frac{36\,mga}{5}\frac{5}{12a}$$

$$= 3\,mg$$

that is,

$$\frac{3T_1}{5} + \frac{4T_2}{5} = 3\,mg,$$

that is,

$$3T_1 + 4T_2 = 15\,mg. \quad (1)$$

Resolving vertically,

$$T_1 \cos \theta - T_2 \sin \theta = mg$$

that is,

$$4T_1 - 3T_2 = 5\,mg. \quad (2)$$

$(1) \times 4: \qquad 12T_1 + 16T_2 = 60\,mg$

$(2) \times 3: \qquad 12T_1 - 9T_2 = 15\,mg$

Subtracting,

$$25T_2 = 45\,mg$$

$$\therefore \ T_2 = 9\,mg/5.$$

Exercise 11a

All strings are light, and all are inextensible unless otherwise stated.

1   The ends of a 50 cm string are attached to a 2 kg particle and a fixed point on a smooth horizontal table. The particle rotates on the table at such a speed that the tension in the string is 5 N. Find this speed.

2   A string passing through a hole in a smooth horizontal table is attached to a 2 kg particle which hangs vertically and a 4 kg particle which rotates on the table at 30 revolutions per minute. Find the length of the horizontal part of the string.

3   A turntable rotates at 20 revolutions per minute. Find to the nearest cm the maximum distance from the centre at which a particle can be placed without slipping if μ for the particle and the turntable is $\frac{1}{3}$.

4   The ends of an elastic string of natural length 60 cm and modulus 180 N are attached to a 3 kg particle and a point on a smooth horizontal table. Find the extension of the string when the particle is rotating on the table at 4 rad s^{-1}.

5   A conical pendulum of length 75 cm rotates at 4 rad s^{-1}. Find its inclination to the vertical.

6   An elastic string of natural length 80 cm is attached at one end to a point on a smooth horizontal table, and at the other to a 4 kg particle. Every minute the particle describes 20 horizontal circles of radius 90 cm. Find the extension of the string when the particle hangs vertically.

7   A particle describes horizontal circles around the inside of a smooth hemispherical bowl of radius 10 cm which is fixed with its rim horizontal. If the reaction of the bowl on the particle is $1\frac{2}{3}$ times the particle's weight, what is the particle's speed?

8   A conical pendulum consists of a string of length 50 cm which carries a particle of mass 2 kg. At what speed is the tension 40 N?

9   The ends of an elastic string of natural length 60 cm are attached to a point on a smooth horizontal table and to a particle which rotates on the table about this point. When each revolution takes $1\frac{1}{2}$ s the radius is 70 cm. What is the radius when each revolution takes 1 s?

10   Prove that the vertical height of a conical pendulum depends only on its angular speed. Find the vertical height of a conical pendulum which takes $\frac{1}{2}$ s for each revolution.

11   A string passes through a hole in a smooth table. It carries at one end a 2 kg particle which rotates as a conical pendulum, and at the other a 4 kg particle which describes horizontal circles on the table. If the ratio of the angular speeds is 3:1, what is the ratio of the distances from the two particles to the hole?

12   A smooth hemispherical bowl of radius $2a$ is fixed with its rim horizontal, and a particle of mass m describes horizontal circles on its inner surface at a depth of a below the rim. Express the speed and the normal reaction in terms of a, m and g.

13   Particles of equal mass are attached to the mid-point and to one end of a string and the other end is fixed to a point on a smooth horizontal table. When the particles describe horizontal circles prove that the ratio of the tensions in the two sections of string is 3:2.

14   A conical pendulum consists of a 750 g particle attached to the end of an elastic string of natural length 50 cm and modulus 200 N. Find the extension when the angular speed is 6 rad s^{-1}.

15   A string of length 70 cm is fixed at its ends to two points 50 cm apart in a vertical straight line. The string passes through a smooth ring which rotates in a horizontal circle. What is the speed of the ring when the two sections of string meet at right angles?

16   One end of a string of length 80 cm is attached to a point on a smooth horizontal table. The other end is attached to a particle of mass m, and a particle of mass $2m$ is attached at another point. Both particles describe horizontal circles on the table. If the ratio of the tensions in the two sections of string is 3:2, what is the distance from the heavier particle to the fixed point?

17   A particle is attached to one end of an elastic string of natural length 80 cm. When the string is held vertically the particle stretches it to a length of 1 m. What will be the extension when the particle describes horizontal circles about the free end at 2 rad s^{-1} on a smooth table?

18   A string ABC of length 60 cm carries a 500 g particle at its mid-point B, and a 1 kg ring at its lower end C. The end A is attached to a point on a fixed smooth vertical rod which passes through the ring. If B moves in horizontal circles, what must be its angular speed for C to be supported a distance 40 cm below A?

19 A string of length 80 cm is fixed at its ends to two points 40 cm apart in a vertical straight line. A 2 kg particle is attached to the string at its mid-point. Given that the string breaks when its tension is 60 N, find the maximum and the minimum speeds at which the particle can rotate in a horizontal circle with both sections of the string straight.

20 A string of length 120 cm is attached at its ends to two points 80 cm apart in a vertical line. The string passes through a smooth ring of mass m which rotates in a horizontal circle at such a speed that it is 70 cm from the upper end of the string. Prove that the tension in the string is $7\,mg/2$ and the angular speed is $\sqrt{12g}$.

Circular motion of a car or a train

There is just one important difference between the cases of a car and a train: in the former case friction supplies the centripetal force, while in the latter the outer rail performs this task. This means that a car which travels in a circle at too great a speed may either slide or topple, while a train can only topple.

Consider first the case of a vehicle travelling on level ground. If it is assumed that the centre of gravity is midway between the wheels, the relevant forces are as shown in Fig. 11.10, in which it may be supposed that the vehicle is being viewed from behind.

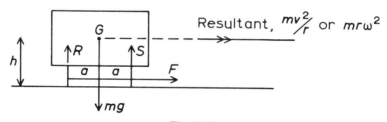

Fig. 11.10

The centre of the circle is somewhere to the right of the diagram, the distance apart of the wheels is $2a$, and the height of the centre of gravity G is h. We shall here accept without proof the fact that the resultant of a system of forces tends to rotate a body if it does not act through the centre of gravity. Since the above body is not rotating the resultant acts through G.

The system of forces shown is not in equilibrium, and we therefore must investigate it by means of the two principles used in Chapters 4, 6 and 7, namely

moment of resultant about any point or axis = moment of original system,

component of resultant in any direction = sum of components of original system.

It follows at once from the latter principle that

$$F = \frac{mv^2}{r} = mr\omega^2$$

and

$$R + S = mg.$$

Clearly the faster the vehicle travels the closer it comes to sliding (if this is possible) or toppling. We shall now derive expressions for the speeds at which these events are just about to occur.

(a) Sliding

The vehicle is on the verge of sliding when $F = \mu(R + S)$, that is when

$$mv^2/r = \mu mg$$

or

$$v = \sqrt{\mu gr}.$$

(b) Toppling

The vehicle is on the point of toppling about its outer wheels when $S = 0$. Thus, using the principle

$$\text{moment of resultant} = \text{moment of original system},$$

we have, taking the axis through the points of contact of the outer wheels,

$$\frac{mv^2 h}{r} = mga$$

$$\therefore v = \sqrt{\frac{gar}{h}}.$$

Consider now the case of a vehicle on a banked road or track. The advantage of banking is that the normal reactions on the wheels have components towards the centre and can thus supply some or all of the centripetal force. The ideal angle of banking for a given speed is that at which all the centripetal force is provided in this way, so that the force F is zero. The force diagram is then as shown in Fig. 11.11.

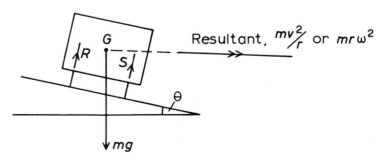

Fig. 11.11

This system of forces is in fact identical in all essential respects to that of the conical pendulum, and we treat it in a similar way.

Resolving horizontally,

$$(R + S)\sin\theta = mv^2/r \quad \text{or} \quad mr\omega^2. \quad (1)$$

Resolving vertically,

$$(R + S) \cos \theta = mg. \quad (2)$$

Dividing (1) by (2):

$$\tan \theta = \frac{v^2}{rg} = \frac{r\omega^2}{g}.$$

Either of these latter two equations gives the ideal banking angle. It is also worth noting that since both the force mg and the resultant pass through G, the resultant of R and S must also pass through this point. Hence, by symmetry, $R = S$.

Example A car travels in a horizontal circle of radius 50 m on a track banked at arc tan $\frac{3}{4}$ to the horizontal. $\mu = \frac{1}{4}$. Find (a) the speed at which there is no lateral force between the wheels and the ground; (b) the minimum speed the car can have without sliding downwards.

(a) Using the result just obtained for the ideal banking angle at a given speed, we have

$$\tan \theta = \frac{v^2}{rg}$$

that is,

$$\frac{3}{4} = \frac{v^2}{500}$$

from which $v = 19.36 \text{ m s}^{-1}$.

(b) When the car is on the point of sliding downwards the frictional force is μ times the total normal reaction and it acts up the slope (Fig. 11.12).

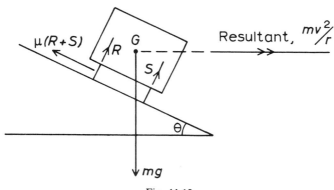

Fig. 11.12

It is convenient here to let $R + S = P$. Then we proceed as follows.

Resolving horizontally,

$$P \sin \theta - \mu P \cos \theta = \frac{mv^2}{r}$$

that is,

$$\frac{3P}{5} - \frac{1}{4}\frac{4P}{5} = \frac{mv^2}{50}$$

that is,
$$\frac{2P}{5} = \frac{mv^2}{50}$$

or
$$P = \frac{mv^2}{20} \quad (1)$$

Resolving vertically,
$$P \cos \theta + \mu P \sin \theta = mg$$

that is,
$$\frac{4P}{5} + \frac{1}{4}\frac{3P}{5} = mg$$

that is,
$$\frac{19P}{20} = mg$$

or
$$P = \frac{20\,mg}{19}. \quad (2)$$

Now from (1) and (2) we have
$$\frac{mv^2}{20} = \frac{20\,mg}{19}$$

from which $v = 14.51$ m s^{-1}.

Exercise 11b

Centres of gravity of cars and trains may be assumed to be midway between the wheels.

1   A 240 tonne train travels at 20 m s^{-1} in a horizontal circle of radius 120 m. Find the lateral force provided by the rails. What banking angle is required to reduce this force to zero?

2   A locomotive's maximum speed round a level circular track of radius 60 m is 19 m s^{-1}. If the rails are 1.4 m apart, what is the height of the centre of gravity to the nearest cm?

3   Trains run at an average speed of 25 m s^{-1} on a track which is part of a circle of radius 750 m. If the rails are 1.5 m apart, how much should the outer rail be raised above the inner to provide ideal banking?

4   The wheels of a car are 1.5 m apart. It is found that the car is on the point of overturning about its outer wheels when it is travelling on a certain level road at 72 km h^{-1} in a circle of radius 80 m. Find (a) the height of the centre of gravity, (b) the minimum possible value of μ.

5   A 2 tonne car runs on a level circular road of radius 150 m. Its wheels are 1.6 m apart and its centre of gravity is 90 cm above the ground. Find μ if it is on the point of sliding and toppling at the same speed.

6   A car travels in a horizontal circle of radius 75 m on a road banked at $20°$, and $\mu = \frac{1}{2}$. At what speed is the car on the point of sliding up the slope, assuming that it does not topple?

7 The wheels of a car are 1.4 m apart and its centre of gravity is 60 cm above the ground. Assuming that it does not slide, find the greatest speed at which the car can travel without overturning on a level road in-a circle of radius 120 m.

8 Repeat question **7**, but now let the road be banked at arc tan $\frac{3}{4}$ to the horizontal.

9 Find the minimum speed at which a car can travel without sliding in a horizontal circle of radius 80 m on a road banked at 30°, if $\mu = 0.3$.

10 A locomotive travels at 20 m s^{-1} in a horizontal circle of radius 96 m. Find the angle at which the track is banked if no lateral force is provided by the rails. Find also the maximum speed of the locomotive on this track, if the rails are 1.5 m apart and the height of the centre of gravity is 1.4 m.

12

Relative Velocity

All velocities are really relative velocities. If we describe a train moving at '80 km h^{-1} due north', we mean that the train has this velocity *relative to the earth*. Most of the velocities which arise in everyday life are relative to the earth, though not all are. A man running at 5 m s^{-1} across the deck of a liner has this velocity relative to the liner, and his velocity relative to the earth is unlikely to be of any interest. Again if we want to talk about the velocity of the earth itself we must find some new reference object or system and measure the earth's velocity relative to this.

It is clear that whenever velocities are specified relative to some reference object, it is convenient to think of that object itself as having a velocity of zero. Now any object can be considered to have a velocity of zero, even though it is usual to take the earth as the basic stationary object. The method of calculating the velocity of one body relative to another is based on the idea of reducing the second body to rest.

Suppose that two bodies A and B have velocities relative to the earth (which we shall henceforth describe for convenience as *true* velocities) of $\mathbf{v}_A$ and $\mathbf{v}_B$ (Fig. 12.1).

Fig. 12.1

If the velocity of A relative to B is required, B must be reduced to rest. We therefore add (vectorially) a velocity of $-\mathbf{v}_B$ to each of the original velocities (Fig. 12.2). This does not affect the relative motion.

Fig. 12.2

B is now at rest and *A* has a velocity of $\mathbf{v}_A - \mathbf{v}_B$. It follows that this is the velocity of *A* relative to *B*, and we thus have the general result:

> **velocity of *A* relative to *B* = velocity of *A* − velocity of *B***

The velocity of *A* relative to *B* can also be described as the velocity which *A* *appears* to have, to an observer moving with *B*. If this idea is employed it must be realised that the apparent velocity means the velocity which an observer calculates on the basis that he himself is at rest.

Worked examples

Preliminary points

1. All relative-velocity problems involve addition and subtraction of vectors. This can either be done by drawing a triangle or parallelogram of velocities, or by use of the unit vectors **i** and **j**. The two methods are not always equally suitable and judgement as to which to use is required in each individual case. The triangle method is probably the better for most simple problems.
2. If a triangle of velocities is drawn *each side should be fully labelled*. A useful method is to put the vector expression ($\mathbf{v}_A, \mathbf{v}_B, \mathbf{v}_A - \mathbf{v}_B$, etc.) first, followed by the magnitude of the vector in brackets.
3. Many problems require a *position diagram* (or space diagram) to be drawn in addition to the velocity diagram.
4. It should be noted that when a boat sails through moving water, or an aeroplane flies through moving air, the direction *aimed* at (i.e. the direction of the course set) is the direction of the *relative* velocity with respect to the water or air. (The true velocity is the relative velocity added to that of the water or air.)

Example 1 A man running due north at $8\,\mathrm{m\,s^{-1}}$ experiences a wind which appears to come from the north-east at $10\,\mathrm{m\,s^{-1}}$. Find the true velocity of the wind.

Method (a) Let the velocity of the man be $\mathbf{v}_M$ and the true velocity of the wind be $\mathbf{v}_W$; then the velocity of the wind relative to the man is $\mathbf{v}_W - \mathbf{v}_M$. The magnitudes and directions of the two given vectors are shown in Fig. 12.3.

Fig. 12.3

A triangle of velocities can now be constructed by using the fact that

$$(\mathbf{v}_W - \mathbf{v}_M) + \mathbf{v}_M = \mathbf{v}_W,$$

and this is shown in Fig. 12.4.

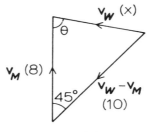

Fig. 12.4

By the cosine formula,

$$x^2 = 10^2 + 8^2 - 2 \times 10 \times 8 \cos 45°$$
$$= 164 - 160 \cos 45°$$
$$= 50.86$$
$$\therefore x = 7.132.$$

Now by the sine formula,

$$\frac{\sin \theta}{10} = \frac{\sin 45°}{7.132},$$

from which $\theta = 82.51°$.

The velocity of the wind is thus **7.132 m s^{-1} in a direction 82.51° W of N** (or on a bearing **277.49°**)

Method (b) Let the unit vectors **i** and **j** be directed towards the east and the north, respectively. Then we have

$$\mathbf{v}_M = 8\mathbf{j},$$
$$\mathbf{v}_W - \mathbf{v}_M = -10 \cos 45° \mathbf{i} - 10 \cos 45° \mathbf{j}.$$

Now since

$$\mathbf{v}_W = (\mathbf{v}_W - \mathbf{v}_M) + \mathbf{v}_M,$$

we have

$$\mathbf{v}_M = -10 \cos 45° \mathbf{i} + (8 - 10 \cos 45°)\mathbf{j},$$

as shown in Fig. 12.5.

Fig. 12.5

By Pythagoras

$$x^2 = 100 \cos^2 45° + (8 - 10 \cos 45°)^2$$
$$= 50.86$$
$$\therefore x = 7.132.$$

Also

$$\tan \theta = \frac{10 \cos 45°}{8 - 10 \cos 45°}$$

from which $\theta = 82.51°$.

In this case there is little to choose between the two methods.

Example 2 Two cars, A and B, are moving at $80\,\mathrm{km\,h^{-1}}$ and $50\,\mathrm{km\,h^{-1}}$, respectively, along two perpendicular roads. Both cars are approaching the intersection of the roads and at a certain moment A is 2 km, and B 3 km, from this point. Find (a) the velocity of A relative to B, (b) the shortest distance apart of the cars.

The position diagram is as shown in Fig. 12.6.

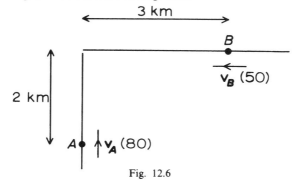

Fig. 12.6

(a) The simplest way to obtain $\mathbf{v}_A - \mathbf{v}_B$ is to add $-\mathbf{v}_B$ to $\mathbf{v}_A$ (Fig. 12.7).

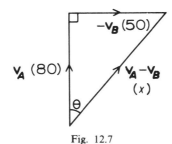

Fig. 12.7

By Pythagoras,
$$x^2 = 80^2 + 50^2$$
$$= 8900$$
$$\therefore\ x = 94.34.$$
Also
$$\tan\theta = 5/8$$
$$\therefore\ \theta = 32°.$$

The velocity of A relative to B is thus **$94.34\,\mathrm{km\,h^{-1}}$ at $32°$ to the direction of A**.

(b) We now return to the position diagram and regard B as at rest; A can then be considered to move with the relative velocity just found (Fig. 12.8).
The shortest distance apart of the cars is clearly SR. Now
$$PQ = 2\tan 32° = 2 \times \tfrac{5}{8} = \tfrac{5}{4}$$
$$\therefore\ QR = 3 - \tfrac{5}{4} = \tfrac{7}{4}$$
$$\therefore\ SR = \tfrac{7}{4}\cos 32°$$
$$= \mathbf{1.484\,km}.$$

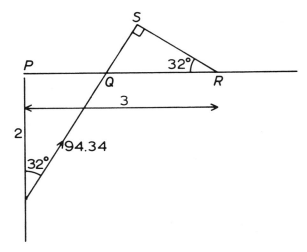

Fig. 12.8

Example 3 A man wishes to swim across a river 50 m wide which is flowing at $0.5\,\mathrm{m\,s^{-1}}$. Given that he can swim at $0.8\,\mathrm{m\,s^{-1}}$, find (a) the time he takes if he decides to swim straight across the river, (b) the shortest possible time in which he can reach the opposite bank.

Let $\mathbf{v}_M$ and $\mathbf{v}_R$ be the true velocities of the man and the river. The $0.8\,\mathrm{m\,s^{-1}}$ is clearly the man's speed relative to the water, and it is thus the magnitude of the quantity $\mathbf{v}_M - \mathbf{v}_R$.

(a) If the man's resultant velocity is straight across the river we have the triangle of velocities shown in Fig. 12.9.

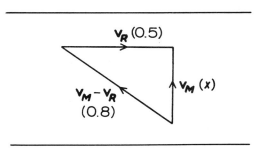

Fig. 12.9

Letting x be the magnitude of the man s true velocity, we obtain by Pythagoras

$$x^2 = 0.8^2 - 0.5^2$$
$$= 0.39$$
$$\therefore\ x = 0.6245\,\mathrm{m\,s^{-1}}.$$

For a body travelling at constant speed we have
$$\text{time} = \text{distance/speed};$$
hence time taken to cross river $= 50/0.6245$
$$= \textbf{80.06 s}.$$
(b) To reach the opposite bank in the shortest possible time the man must proceed just as he would if the water were stationary, i.e. *aim* straight at the opposite bank. The movement of the water merely adds a component which carries him downstream, and does not affect his velocity perpendicular to the banks. We thus have:
$$\text{component perpendicular to banks} = \text{relative velocity}$$
$$= 0.8.$$
Hence
$$\text{time} = \text{distance/speed} = 50/0.8$$
$$= \textbf{62.5 s}.$$

Example 4 The pilot of an aeroplane wishes to fly on a bearing of $60°$, but a 100 km h^{-1} wind is blowing on a bearing of $120°$. Find (a) the course he should set if his speed relative to the air is to be 250 km h^{-1}, (b) the minimum speed relative to the air at which he can achieve the desired direction at all.

Let the true velocities of plane and wind be v_P and v_W, respectively; then $v_P - v_W$ is the relative velocity.
(a) The velocities of the plane and of the wind, respectively, are shown in Figs. 12.10 and 12.11.

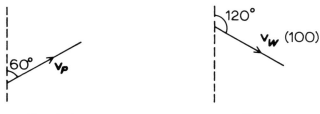

Fig. 12.10   Fig. 12.11

The triangle of velocities (Fig. 12.12) can be obtained from these by adding $-v_W$ to v_P to get $v_P - v_W$.

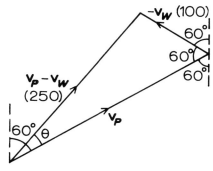

Fig. 12.12

By the sine formula,

$$\frac{\sin\theta}{100} = \frac{\sin 60°}{250}$$

from which $\theta = 20.27°$.

The course set (i.e. the direction of the relative velocity) should be $60° - 20.27°$ = **39.73°**.

(b) In the above triangle of velocities the magnitude and direction of $-\mathbf{v}_W$ are fixed, and the direction of $\mathbf{v}_P$ is fixed. The third side, representing $\mathbf{v}_P - \mathbf{v}_W$, is variable in both magnitude and direction, and it has its least magnitude when it is perpendicular to $\mathbf{v}_P$ (Fig. 12.13). Clearly the value of this magnitude is $100\sin 60°$, i.e. 86.6; hence the least speed at which the desired direction of flight is possible is **86.6 km h^{-1}**.

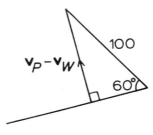

Fig. 12.13

Example 5   A boy P starts to run north-east at $4\,\text{m s}^{-1}$, and at the same time another boy Q, who can run at $6\,\text{m s}^{-1}$, starts to chase him. If Q is initially 10 m due west of P, find the direction he should take in order to catch him as soon as possible, and the distances both boys then run before P is caught.

The initial situation is as shown in Fig. 12.14. If Q is to intercept P as quickly as possible he must run in such a way that his relative velocity is always directed towards P. That is, the relative velocity $\mathbf{v}_Q - \mathbf{v}_P$ must be directed along AB. (This is also clear if we reduce P to rest by adding $-\mathbf{v}_P$ to both velocities.) The triangle of velocities is thus as shown in Fig. 12.15.

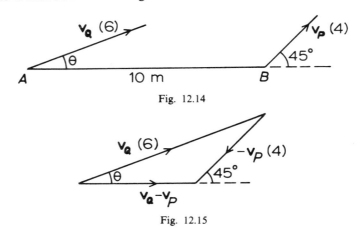

Fig. 12.14

Fig. 12.15

By the sine formula,

$$\frac{\sin\theta}{4} = \frac{\sin 45°}{6} \qquad (\text{since } \sin 135° = \sin 45°)$$

from which

$$\theta = 28.13°.$$

The boy Q should therefore run on a bearing $90° - 28.13°$, i.e. **61.87°**.

To find the distances travelled by the boys we note that since distance is proportional to speed for a body moving at constant speed, any triangle similar to the above triangle of velocities represents the distances travelled by P and Q, and the relative distance, in any given time. When the relative distance is 10 m, we thus have the triangle shown in Fig. 12.16, in which x and y represent the distances travelled by P and by Q.

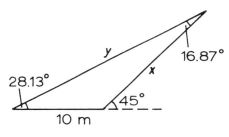

Fig. 12.16

By the sine formula,

$$\frac{x}{\sin 28.13°} = \frac{10}{\sin 16.87°},$$

from which

$$x = 16.25\,\text{m}.$$

Similarly we have

$$y = 24.37\,\text{m}.$$

P and Q thus travel **16.25 m** and **24.37 m**, respectively, before P is caught.

Example 6 (In this example the units of distance and velocity are metres and $m\,s^{-1}$, respectively.) A particle A starts at the point with position vector 12i and moves with a constant velocity of $3i + 5j$, while a particle B is at the point with position vector 6j and moves with a constant velocity of $7i + kj$. Write down an expression for the velocity of B relative to A and hence find the values of k such that (a) the particles collide, (b) B appears to A to be moving parallel to the x-axis. In the latter case find also (c) the time interval for which the particles are within 10 m of each other.

The velocity of B relative to A is $v_B - v_A$, that is

$$7i + kj - (3i + 5j)$$
$$= 4i + (k - 5)j.$$

(a) If the particles are to collide the velocity of B relative to A must be directed towards A (Fig. 12.17).

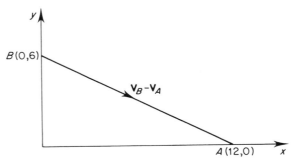

Fig. 12.17

Clearly $\mathbf{v}_B - \mathbf{v}_A$ has to be in the direction of the vector $2\mathbf{i} - \mathbf{j}$ (or $4\mathbf{i} - 2\mathbf{j}$, etc.). Since $\mathbf{v}_B - \mathbf{v}_A = 4\mathbf{i} + (k - 5)\mathbf{j}$, it follows that

$$k - 5 = -2$$
$$\therefore\ k = 3.$$

(b) If the relative velocity is parallel to the x-axis the coefficient of $\mathbf{j}$ must be zero. Hence
$$k - 5 = 0$$
$$\therefore\ k = 5.$$

(c) Regarding A as at rest, B can be considered to move with the relative velocity $4\mathbf{i}$. Fig. 12.18 then shows that the particles are within $10\,\mathrm{m}$ of each other while B travels a distance of $16\,\mathrm{m}$ relative to A.

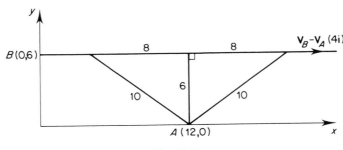

Fig. 12.18

It follows that the required time interval is $16/4 = \mathbf{4\,s}$.

Exercise 12

All the given velocities may be assumed to be constant. When distances or velocities are given in terms of $\mathbf{i}$ and $\mathbf{j}$, the units are metres and $\mathrm{m\,s}^{-1}$, respectively.

1   One woman walks due north at $5\,\mathrm{km\,h}^{-1}$, and another walks south-west at $4\,\mathrm{km\,h}^{-1}$. Find the velocity of the first woman relative to the second (in both magnitude and direction).

2 Two cars move at $24\,\mathrm{m\,s^{-1}}$ and $18\,\mathrm{m\,s^{-1}}$ in the directions north-east and north-west, respectively. Find the velocity of the first car relative to the second.

3 To a man moving due west at $10\,\mathrm{km\,h^{-1}}$ the wind appears to be coming from the north at $15\,\mathrm{km\,h^{-1}}$. Find the true velocity of the wind.

4 The velocity of a particle A is $4\mathbf{i}-3\mathbf{j}$ and that of a particle B relative to A is $\mathbf{i}+2\mathbf{j}$. Find the velocity of B.

5 The velocity of a particle A is $-9\mathbf{i}+4\mathbf{j}$ and that of A relative to a particle B is $-7\mathbf{i}+\mathbf{j}$. Find the velocity of B.

6 To a ship sailing south-west at $10\,\mathrm{m\,s^{-1}}$ another ship appears to be sailing on a bearing of $300°$ at $20\,\mathrm{m\,s^{-1}}$. Find the true velocity of the second ship.

7 A moving body appears to an observer to be travelling due east at $50\,\mathrm{m\,s^{-1}}$. If the body is actually travelling due north at $120\,\mathrm{m\,s^{-1}}$ what is the velocity of the observer?

8 A motor-boat which can travel at $6\,\mathrm{m\,s^{-1}}$ in still water crosses a river $45\,\mathrm{m}$ wide which is flowing at $2\,\mathrm{m\,s^{-1}}$. If the boat is pointed at right angles to the banks find (a) its true speed, (b) the distance it is carried downstream.

9 An aeroplane which can fly at $600\,\mathrm{km\,h^{-1}}$ needs to make a north-easterly journey. If a wind of $100\,\mathrm{km\,h^{-1}}$ is blowing from due east, what course must the pilot set?

10 A river $20\,\mathrm{m}$ wide flows at $3\,\mathrm{m\,s^{-1}}$. If the least time in which a certain boat can cross it is $4\,\mathrm{s}$, how long will the boat take if it travels at right angles to the banks?

11 Ship P is initially $500\,\mathrm{m}$ due west of ship Q. Given that P sails north-east at $15\,\mathrm{m\,s^{-1}}$ and Q sails due north at $8\,\mathrm{m\,s^{-1}}$, find (a) the velocity of P relative to Q, (b) their shortest distance apart.

12 A particle A is initially at the point $(4,-1)$ and a particle B is at the origin. Find the velocity of A relative to B and the shortest distance apart of the particles if the velocities of A and B are (a) $5\mathbf{i}+3\mathbf{j}$ and $5\mathbf{i}-7\mathbf{j}$, respectively, (b) $6\mathbf{i}-2\mathbf{j}$ and $7\mathbf{i}-3\mathbf{j}$, respectively.

13 An aeroplane which can fly at $250\,\mathrm{km\,h^{-1}}$ travels $500\,\mathrm{km}$ due east and then returns directly to its starting point. How long does the complete trip take if an $80\,\mathrm{km\,h^{-1}}$ wind is blowing from due north?

14 A man swims directly across a river which is flowing at $30\,\mathrm{cm\,s^{-1}}$. If he has to direct his body at $40°$ to the banks to achieve this, what is his speed relative to the water?

15 A ship P, sailing due north at $20\,\mathrm{km\,h^{-1}}$, sights a ship Q $2\,\mathrm{km}$ away on a bearing of $90°$. From P, Q appears to be sailing north-west at $5\,\mathrm{km\,h^{-1}}$. Find the distance apart of the ships when they are closest together and the distance travelled by P when this point is reached.

16 A particle P starts at the point with position vector $6\mathbf{i}$ and moves with a velocity of $9\mathbf{i}+3\mathbf{j}$, while a particle Q starts simultaneously at the point with position vector $12\mathbf{j}$ and moves with a velocity of $3\mathbf{i}-5\mathbf{j}$. Find the minimum distance between the particles and the time interval for which they are less than $13\,\mathrm{m}$ apart.

17 Two cars, travelling at $20\,\mathrm{m\,s^{-1}}$ and $15\,\mathrm{m\,s^{-1}}$, approach the intersection of two perpendicular roads. At a certain moment both cars are $300\,\mathrm{m}$ from the intersection. Find the distance of the second car from the intersection when the cars are at their minimum distance apart.

18 Ship P is initially $5\,\mathrm{km}$ due north of ship Q. If P sails at $40\,\mathrm{km\,h^{-1}}$ on a bearing of $120°$ and Q sails at $30\,\mathrm{km\,h^{-1}}$ on a bearing of $30°$, how far apart are the ships when they are closest together and how long does it take them to reach this position?

19   A particle A starts at the origin and moves with a velocity of $5\mathbf{i} + 7\mathbf{j}$, while a particle B starts simultaneously at the point $(4,2)$ and moves with a velocity of $3\mathbf{i} + k\mathbf{j}$. Find the value of k such that (a) the particles collide, (b) B appears to A to be moving parallel to the x-axis, (c) A appears to B to be moving in the direction of the line $y = x$.

20   A body P starts to move at $20\,\mathrm{m\,s^{-1}}$ due east, and at the same time a body Q, initially 60 m due south of P, sets off at $30\,\mathrm{m\,s^{-1}}$. Given that the bodies eventually collide, find the direction in which Q moves and the distance travelled by each body before the collision.

21   A body P starts to move at $8\,\mathrm{m\,s^{-1}}$ on a bearing of $135°$, and at the same time a body Q, initially 20 m due west of P, sets off. Given that the bodies collide after 4 s, find the velocity of Q.

22   John runs at $5\,\mathrm{m\,s^{-1}}$ on a bearing of $30°$, while Harry runs at $8\,\mathrm{m\,s^{-1}}$ on a bearing of $120°$. Given that they meet after 6 s, find the initial distance and bearing of John from Harry.

23   A woman swims across a river 40 m wide which is flowing at $2\,\mathrm{m\,s^{-1}}$, reaching a point 30 m downstream from her original position. If she swims at twice the minimum speed (relative to the water) which would get her to this point, how long does the journey take?

24   An aeroplane with a speed in still air of $200\,\mathrm{km\,h^{-1}}$ flies north-east for 300 km, then south-east for the same distance. Given that a wind of $120\,\mathrm{km\,h^{-1}}$ blows from the south throughout, find the total time of the trip to the nearest minute.

25   To an observer who is moving south-east at $5\,\mathrm{m\,s^{-1}}$, a certain body appears to be moving due west. Given that their minimum distance apart is 48 m, and that they are within 50 m of each other for 4 s, find the true velocity of the body.

26   A ship needs to sail due north in a current of $40\,\mathrm{km\,h^{-1}}$ coming from $30°$ east of south. What course should the pilot set in order that the speed relative to the water should be as small as possible, and what will be the ship's relative speed and its true speed in this case?

27   Ann runs at $5\,\mathrm{m\,s^{-1}}$ on a bearing of $300°$ and at the same time Jane, initially due west of Ann, sets off in an attempt to catch her. Find the minimum speed of Jane if she is to succeed.

28   An observer moving at $4\,\mathrm{m\,s^{-1}}$ sights a body which is in fact moving at $6\,\mathrm{m\,s^{-1}}$ due north. If the body appears to the observer to be moving on a bearing of $30°$, what are the two possible directions of the observer's motion?

29   A viewer on a ship sailing at $30\,\mathrm{km\,h^{-1}}$ observes that another ship to the north-west appears to be moving straight towards him at $50\,\mathrm{km\,h^{-1}}$. What is the minimum possible bearing on which the second ship is sailing?

30   Bodies A and B move at $16\,\mathrm{m\,s^{-1}}$ due east and $12\,\mathrm{m\,s^{-1}}$ due south, respectively. After 3 s they reach the position at which they are their shortest distance apart, namely 25 m. Find the original distance and the two possible values of the original bearing of B from A.

13

Projectiles

In this chapter we consider the motion of a particle which is projected at any angle to the horizontal and then moves under the influence of its own weight alone. The effect of air friction is thus ignored. Since the only force on such a particle is its own weight, which always acts vertically downwards, the acceleration of the particle has a constant downward component of g and a horizontal component of zero. The whole motion of a projectile can in fact be considered to consist of two superimposed component motions, namely

(a) a pure vertical motion with a downward acceleration of g;
(b) a pure horizontal motion with constant speed.

Motion (a) is governed by the five equations considered in Chapter 1, while motion (b) is governed by the single equation $s = ut$. The effect of superimposing the motions is to produce a path which will shortly be shown to be a parabola.

It is clear that the complete motion of a projectile is determined when its initial velocity is given. The initial velocity is normally specified in one of two ways: by stating its magnitude and inclination to the horizontal, or by using the unit vectors **i** and **j**. It is usually convenient to take a pair of x and y axes as shown in Figs. 13.1 and 13.2, which illustrate the two ways of specifying the initial velocity.

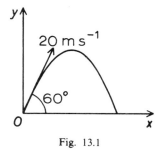

Fig. 13.1

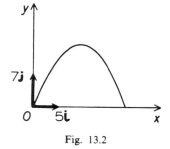

Fig. 13.2

Many simple problems on projectiles can be solved with no more theory than that just outlined. Either the horizontal or vertical motion alone is considered, and the appropriate equation applied. The method is illustrated by the following examples.

Example 1 A ball is thrown from a point O with a speed of $20\,\text{m}\,\text{s}^{-1}$ at an elevation of $40°$. Find (a) the magnitude and direction of its velocity, (b) its distance from O, after $\frac{1}{2}\,\text{s}$.

Let the ball be at the point $P(x, y)$ $\frac{1}{2}$s after leaving O. Figure 13.3 shows the relevant components of velocity. Note that the horizontal component, being constant, is the same at O and P.

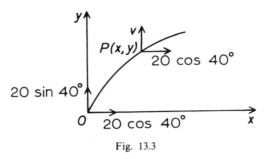

Fig. 13.3

(a) To find v we consider the vertical component of the motion alone. Using

$$v = u + at,$$

we have

$$v = 20 \sin 40° - 10(\tfrac{1}{2})$$
$$7.856.$$

Now letting the speed at P be $w\,\mathrm{m\,s^{-1}}$, and the elevation θ, we have Fig. 13.4.

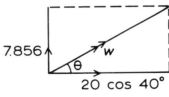

Fig. 13.4

By Pythagoras,

$$w^2 = (20 \cos 40°)^2 + 7.856^2$$
$$= 296.45$$
$$\therefore\ w = 17.22$$

Also

$$\tan \theta = \frac{7.856}{20 \cos 40°}$$

from which

$$\theta = 27.15°.$$

The velocity at P is thus **$17.22\,\mathrm{m\,s^{-1}}$ at an elevation of $27.15°$**.

(b) To find the distance OP we require x and y. Considering the horizontal motion, we have

$$x = 20 \cos 40° \times \tfrac{1}{2}$$
$$= 7.660.$$

Considering the vertical motion and using

$$s = ut + \tfrac{1}{2}at^2,$$

we have

$$y = (20 \sin 40°)(\tfrac{1}{2}) - 5(\tfrac{1}{2})^2$$
$$= 5.178.$$

Hence

$$OP^2 = 7.660^2 + 5.718^2,$$

from which

$$OP = 9.559 \text{ m}.$$

Example 2 A body is thrown straight out to sea from the top of a cliff with an initial velocity, in m s^{-1}, given by $6\mathbf{i} + 4\mathbf{j}$. Given that it strikes the water after 3 s, find (a) the height of the cliff, (b) the horizontal distance of the body from the cliff when it strikes the water.

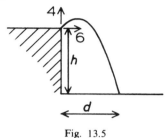

Fig. 13.5

(a) Consider the vertical motion alone. Using

$$s = ut + \tfrac{1}{2}at^2$$

and taking downward as positive, we have

$$h = -(4 \times 3) + 5(3^2)$$
$$= 33 \text{ m}.$$

(b) Consider the horizontal motion alone. Using $s = ut$, we have

$$d = 6 \times 3$$
$$= 18 \text{ m}.$$

A somewhat more difficult example, but still requiring no further theory, is the following.

Example 3 A body is projected with a velocity of $5\mathbf{i} + 10\mathbf{j}$, and 1 s later another body is projected from the same point with a velocity of $u\mathbf{i} + 8\mathbf{j}$. What is the value of u if the bodies collide? (The units of velocity are m s^{-1}.)

First we find the time at which the bodies have the same height. Let both bodies have a height of h at time t after the first is projected. The second body will then have been in motion for a time of $t - 1$.

Consider the vertical motion alone. Using the equation

$$s = ut + \tfrac{1}{2}at^2,$$

we have for the first body

$$h = 10t - 5t^2,$$

and for the second,

$$h = 8(t - 1) - 5(t - 1)^2.$$

Hence
$$10t - 5t^2 = 8t - 8 - 5t^2 + 10t - 5,$$
from which
$$t = \tfrac{13}{8}.$$

Now if the bodies are to collide they must have the same horizontal distance after this time. Hence, considering the horizontal motion alone, we have
$$5 \times \tfrac{13}{8} = u(\tfrac{13}{8} - 1)$$
$$= 5u/8.$$

Hence $u = 13$.

Exercise 13a

When velocities are given in terms of i and j the units are $m\,s^{-1}$.

1   A body is thrown with an initial velocity of $5i + 8j$. Find its velocity in terms of i and j after $2\,s$.

2   A body is projected from a point O with an initial velocity of $6i + 8j$. Find its position vector with respect to O after $\tfrac{1}{2}\,s$.

3   A ball is thrown horizontally from the top of a building $20\,m$ high with a speed of $6\,m\,s^{-1}$. Find its horizontal distance from the building when it hits the ground.

4   A body is projected with a speed of $25\,m\,s^{-1}$ at an elevation of arc tan $\tfrac{4}{3}$. Find its distance from the starting point after $0.4\,s$.

5   A stone is thrown straight out to sea off a cliff $40\,m$ high with a speed of $20\,m\,s^{-1}$ and an elevation of $30°$. Find the time it takes to hit the water and its distance from the foot of the cliff when it does so.

6   A body is projected with a velocity of $12i - 10j$. Find its speed and inclination to the horizontal after $1\,s$.

7   A ball is thrown which just clears a wall $7.2\,m$ high and $12\,m$ away after $0.6\,s$. Find the initial velocity in terms of i and j.

8   A body is thrown horizontally from the top of a building $45\,m$ high, and strikes the ground $45\,m$ from the foot of the building. At what speed is it thrown?

9   A ball is thrown with a speed of $12\,m\,s^{-1}$ at an elevation of $30°$. Find (a) the times at which its height is $175\,cm$, (b) the horizontal distance it travels between these times.

10   After $2\,s$ of flight the velocity of a projectile is $12i - 4j$. Find its speed and elevation at the moment of projection.

11   A ball is thrown from a height of $320\,cm$ with a velocity of $20i + 15j$. At what inclination to the horizontal does it strike the ground?

12   After travelling a horizontal distance of $8\,m$, a projectile has a velocity of $16i + 4j$. Find its initial velocity in terms of i and j.

13   A body is projected with a velocity of $10i + 8j$. Calculate the horizontal distance it travels while more than $3\,m$ above its starting point.

14   A body is projected with a speed of $25\,m\,s^{-1}$ at an elevation of $50°$. What is its height when it is moving horizontally?

15   A body is thrown with an initial velocity of $8i + 12j$. Find the two times at which its inclination to the horizontal is arc tan $\tfrac{1}{2}$.

16   A body is projected from a point O, and after $2\,s$ its position vector with respect to O is $10i - 2j$. Find its initial velocity in terms of i and j.

17 A body is projected with a velocity of $8\mathbf{i} + 12\mathbf{j}$, and 1 s later another body is projected from the same point with a velocity of $6\mathbf{i} + 10\mathbf{j}$. Find the time at which their heights are equal, and their horizontal distance apart at this time.

18 A body is projected with a velocity of $5\mathbf{i} + 8\mathbf{j}$, and $\frac{1}{2}$ s later another body is projected from the same point with a velocity of $u\mathbf{i} + 10\mathbf{j}$. Find u if the bodies collide.

19 At the moment when one boy throws a ball from ground level, another boy 3 m away drops a ball from a point 2 m higher. Given that the two balls collide when they are 20 cm above the ground, find the initial speed of the first ball.

20 A ball is thrown downwards from the top of a tower with a speed of $20 \, \text{m s}^{-1}$ at an angle of arc $\tan \frac{3}{4}$ to the horizontal. If it strikes the ground at an angle of arc $\tan 3$ to the horizontal, find the height of the tower and the distance of the ball from the foot of the tower when it hits the ground.

21 Two boys 10.5 m apart throw balls towards each other from the same height with initial speeds of $15 \, \text{m s}^{-1}$. If the initial elevations are arc $\tan \frac{3}{4}$ and arc $\tan \frac{4}{3}$, how far apart are the balls when they cross?

22 A boy throws a ball from the top of a building 10 m high and aims to hit a point on the ground which is 10 m from the foot of the building. At what speed must he throw it if its initial elevation is 45°?

Standard equations for projectiles

Consider a projectile which starts from a point on a horizontal plane with a speed V at an elevation α, moves freely under gravity, and finally returns to the plane. We shall derive expressions for the *time of flight*, T, the *horizontal range*, R, and the *greatest height*, H. We shall also obtain the equation of the projectile's path, using a pair of axes positioned as in the last sections (Fig. 13.6).

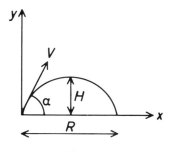

Fig. 13.6

Time of flight, T

Consider the vertical motion alone and let the upward direction be positive. For the complete flight we have

$$u = V \sin \alpha, \quad a = -g, \quad s = 0, \quad t = ?$$

Hence, using

$$s = ut + \tfrac{1}{2}at^2,$$

we have

$$0 = (V\sin\alpha)t - gt^2/2$$
$$= t(V\sin\alpha - gt/2)$$
$$\therefore \ t = 0 \quad \text{or} \quad \frac{2V\sin\alpha}{g}.$$

Clearly $t = 0$ at the start of the flight; hence

$$\boxed{T = \frac{2V\sin\alpha}{g}}$$

Horizontal range, R

Consider the horizontal motion alone. Since $a = 0$, this is governed by the simple equation $s = ut$.

Now $u = V\cos\alpha$, and as just shown, when $s = R$, $t = (2V\sin\alpha)/g$. Hence

$$R = \frac{2V^2\sin\alpha\cos\alpha}{g}$$

or

$$\boxed{R = \frac{V^2\sin 2\alpha}{g}}$$

Greatest height, H

Consider the vertical motion alone. For the first half of the flight we have

$$u = V\sin\alpha, \quad a = -g, \quad s = H, \quad v = 0.$$

Hence, using

$$v^2 = u^2 + 2as,$$

we have

$$0 = V^2\sin^2\alpha - 2gH$$

and thus

$$\boxed{H = \frac{V^2\sin^2\alpha}{2g}}$$

The equation of the path

The task is to obtain an equation relating x and y which is true for any point P on the curve (Fig. 13.7). This equation must involve only x, y and the *constants* V, α and g.

Taking the period from O to P and applying the equation $s = ut + \frac{1}{2}at^2$ to the vertical component of the motion, we have

$$y = (V\sin\alpha)t - gt^2/2. \quad (1)$$

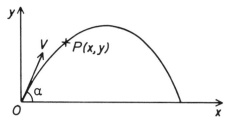

Fig. 13.7

Considering the horizontal component, and applying the equation $s = ut$ to the same period, we obtain

$$x = (V \cos \alpha)t. \quad (2)$$

The variable t must now be eliminated. From (2),

$$t = \frac{x}{V \cos \alpha}.$$

Substituting this value for t in (1), we obtain the equation of the path. Since the right hand side is a quadratic in x, the equation represents a *parabola*.

$$y = x \tan \alpha - \frac{gx^2 \sec^2 \alpha}{2V^2}$$

It is important to note also that since $\sec^2 \alpha = 1 + \tan^2 \alpha$, the equation of the path can be regarded as a *quadratic equation in tan* α. An application of this is given in Example 4 below.

Two corollaries from the result for range

Corollary 1 Since the maximum value of $\sin 2\alpha$ is 1, which occurs when $2\alpha = 90°$, we have the following formula for the maximum horizontal range.

$$\textbf{Maximum range} = \frac{V^2}{g},$$

occurring when $\alpha = 45°$**.**

Corollary 2 For any range below the maximum, there are two possible angles of projection, which add up to 90°.

Proof Since

$$R = \frac{V^2 \sin 2\alpha}{g},$$

$$\sin 2\alpha = \frac{Rg}{V^2}.$$

The angles required to achieve a range of R with an initial speed of V are given by the solutions for α of this equation. Now an equation of this form in general has two solutions which add up to $180°$. Let these be x and $180° - x$. We then have

$$2\alpha = x \qquad \text{or} \qquad 2\alpha = 180° - x$$
$$\therefore \quad \alpha = x/2 \quad \text{or} \qquad \alpha = 90° - x/2$$

There are thus two possible angles, which add up to $90°$.

Worked examples

In many problems it makes little difference whether one of the standard results is used or whether the earlier approach of simply considering vertical and horizontal components is adopted. There are some problems, however, in which the standard results help a good deal, and some examples of this kind will now be given.

Example 1   Find the initial angle of elevation of a projectile whose horizontal range is 5 times its greatest height.

We have

$$\frac{V^2 \sin 2\alpha}{g} = \frac{5 V^2 \sin^2 \alpha}{2g}$$

$$\therefore \quad 2 \sin \alpha \cos \alpha = \frac{5 \sin^2 \alpha}{2}.$$

Dividing both sides by $\sin \alpha \cos \alpha$, we obtain

$$\tan \alpha = \tfrac{4}{5},$$

and thus

$$\alpha = \textbf{arc tan} \tfrac{4}{5}.$$

Example 2   The maximum distance a boy can throw a ball is $40\,\text{m}$. At what elevations would he achieve a distance of $25\,\text{m}$, throwing the ball at the same speed?

We have

$$\frac{V^2}{g} = 40;$$

hence

$$V^2 = 400 \quad \text{and} \quad V = 20.$$

Now using

$$\sin 2\alpha = \frac{Rg}{V^2},$$

and letting $R = 25$, we have

$$\sin 2\alpha = \frac{250}{400}$$

$$\therefore \quad 2\alpha = 38.68° \quad \text{or} \quad 180° - 38.68°$$

and

$$\alpha = 19.34° \quad \text{or} \quad 70.66°.$$

(Note that these angles add up to $90°$.)

Example 3   A ball is thrown from a point on the ground towards a vertical wall which is 6 m away. It strikes the wall at right angles at a height of 4 m, and bounces back to hit the ground at a point 3 m from the wall. Find the ball's initial elevation and speed, and the coefficient of restitution for the ball and the wall (Fig. 13.8).

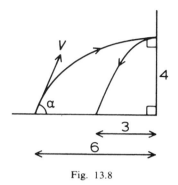

Fig. 13.8

Clearly $R = 12$ and $H = 4$. Hence $R = 3H$ and we have

$$\frac{V^2 \sin 2\alpha}{g} = \frac{3 V^2 \sin^2 \alpha}{2g}$$

$$\therefore \; 2 \sin \alpha \cos \alpha = \frac{3 \sin^2 \alpha}{2}.$$

Dividing both sides by $\sin \alpha \cos \alpha$, we get

$$\tan \alpha = \tfrac{4}{3}$$

and hence

$$\alpha = \arctan \tfrac{4}{3}.$$

It follows immediately that $\sin \alpha = \tfrac{4}{5}$ and $\cos \alpha = \tfrac{3}{5}$, and we can therefore obtain V by substituting these values into the equation for R:

$$R = \frac{V^2 \sin 2\alpha}{g} = \frac{2 V^2 \sin \alpha \cos \alpha}{g}$$

Hence

$$12 = \frac{2 V^2 \times \tfrac{4}{5} \times \tfrac{3}{5}}{10},$$

from which $V = 11.18 \, \text{m s}^{-1}$.

The coefficient of restitution, e, is simply

$$\frac{\text{horizontal speed just after collision}}{\text{horizontal speed just before collision}}.$$

Now both horizontal speeds are constant, and both are given by the expression

$$\frac{\text{horizontal distance travelled}}{\text{time taken}}.$$

By considering the vertical motions, however, we can see that the time for the upward journey equals that of the downward journey. Letting this time be t, therefore, we have

$$e = \frac{3/t}{6/t}$$

$$= \tfrac{1}{2}.$$

Example 4   A ball is thrown at 10 m s^{-1} and aimed at a point which is 4 m away and 4 m above the point of projection. Find the two possible angles of projection, and the complete horizontal range corresponding to the larger of these angles.

Here we use the equation of the path. The information given tells us that the path passes through the point (4,4) (Fig. 13.9).

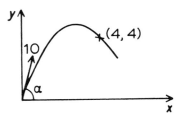

Fig. 13.9

Substituting $x = 4$, $y = 4$, $V = 10$ into the equation of the path, and using the identity $\sec^2 \alpha = 1 + \tan^2 \alpha$, we have

$$4 = 4 \tan \alpha - \frac{160(1 + \tan^2 \alpha)}{200}$$

$\therefore \qquad 800 = 800 \tan \alpha - 160 - 160 \tan^2 \alpha$

$\therefore \quad 160 \tan^2 \alpha - 800 \tan \alpha + 960 = 0$

$\therefore \quad \tan^2 \alpha - 5 \tan \alpha + 6 = 0$

$\therefore \quad (\tan \alpha - 3)(\tan \alpha - 2) = 0$

$\therefore \quad \alpha = \textbf{arc tan 3} \quad \text{or} \quad \textbf{arc tan 2}.$

To find R when $\tan \alpha = 3$ we first obtain $\sin \alpha$ and $\cos \alpha$ by constructing a right angled triangle as shown in Fig. 13.10. Clearly $h = \sqrt{10}$ and hence $\sin \alpha = 3/\sqrt{10}$ and $\cos \alpha = 1/\sqrt{10}$.

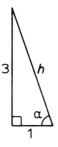

Fig. 13.10

Now we have

$$R = \frac{V^2 \sin 2\alpha}{g}$$

$$= \frac{2V^2 \sin \alpha \cos \alpha}{g}$$

$$= \frac{200 \times 3/\sqrt{10} \times 1/\sqrt{10}}{10}$$

$$= \mathbf{6\,m}.$$

Exercise 13b

1 A body is projected with a speed of $20\,\mathrm{m\,s^{-1}}$. Find the horizontal range if the initial elevation is $60°$, and the maximum horizontal range.

2 A ball thrown at $15\,\mathrm{m\,s^{-1}}$ reaches a maximum height of $8\,\mathrm{m}$. Find the angle of projection.

3 The maximum range of a projectile is $22.5\,\mathrm{m}$. Find its time of flight when achieving this range.

4 Find the horizontal range of a projectile thrown at arc tan $\frac{4}{3}$ to the ground which takes $2\,\mathrm{s}$ to reach its maximum height.

5 A ball thrown at $8\,\mathrm{m\,s^{-1}}$ is aimed at a point on the same level $5\,\mathrm{m}$ away. Find the two angles of projection which will achieve this.

6 Find the minimum speed with which a ball must be thrown if it is to have a horizontal range of $50\,\mathrm{m}$.

7 A body is projected with a speed of $25\,\mathrm{m\,s^{-1}}$ at an elevation of arc tan $\frac{3}{4}$. Find the height of the body above its starting point when it has travelled $20\,\mathrm{m}$ horizontally.

8 The maximum distance a man can throw a ball is $60\,\mathrm{m}$. What horizontal range will he achieve if he throws the ball at maximum speed, but at $30°$ to the ground?

9 If the horizontal range of a projectile is equal to its greatest height, what is the initial angle of elevation?

10 A body projected at $50\,\mathrm{m\,s^{-1}}$ reaches a maximum height of $80\,\mathrm{m}$. Find its horizontal range.

11 Find the horizontal range of a projectile thrown at $20\,\mathrm{m\,s^{-1}}$ whose time of flight is $3\,\mathrm{s}$.

12 What is the minimum speed at which an object must be projected to have a horizontal range of $90\,\mathrm{m}$, and what is the corresponding greatest height?

13 Find the ratio of the times of the two flights for which a projectile has half its maximum possible horizontal range.

14 A ball is thrown at $20\,\mathrm{m\,s^{-1}}$ from a point $5\,\mathrm{m}$ above the ground. Find the two angles of projection which make it reach the ground $40\,\mathrm{m}$ away.

15 A body projected at a certain speed achieves the same horizontal range with angles of projection of α_1 and α_2. Prove that if the corresponding greatest heights are H_1 and H_2, then $H_1/H_2 = \tan^2 \alpha_1$.

16 A ball is thrown at arc tan $\frac{4}{3}$ to the horizontal, and just passes over a wall $15\,\mathrm{m}$ high and $30\,\mathrm{m}$ distant. Find the distance behind the wall of the point at which it reaches its original level.

17   A ball thrown at an elevation of 30° strikes a wall at right angles after $\frac{3}{4}$ s. If the coefficient of restitution is $\frac{2}{3}$, how far in front of its starting point is the ball when it reaches its original level?

18 T_1 and T_2 are the times of the two flights with which a body projected at a certain initial speed achieves a range of R. Prove that $R = gT_1T_2/2$.

19   A ball thrown with a speed of $10\,\text{m}\,\text{s}^{-1}$ just clears a post of height 2.5 m which is 5 m distant. How far is it from the post when it reaches its original level, if it is thrown so as to make this distance as small as possible?

20   A boy throws a ball at a wall from a height of 35 cm. It strikes the wall normally at a height of 80 cm, rebounds, and hits the ground at the boy's feet, directly below the point of projection. Find the coefficient of restitution for the ball and the wall.

21   A ball is thrown so as just to clear two posts 8 m high which are 24 m apart. The ball starts at ground level a distance 12 m from the nearer post. Find the tangent of its angle of projection.

14

Centres of Gravity and Mass, Centroids

The meaning of 'centre of gravity'

Any extended body can be considered to be made up of many particles. Unless the body is very large, the weights of these particles form a system of parallel forces. Consider just three particles A, B, C, with weights w_1, w_2, w_3 (Fig. 14.1). Whatever the orientation of the body relative to the earth, the resultant of w_1 and w_2 is a parallel force of $w_1 + w_2$, which passes through P, where $AP/BP = w_2/w_1$.

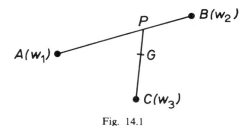

Fig. 14.1

It follows that the resultant of all three forces is a parallel force of

$$w_1 + w_2 + w_3,$$

which passes through G, where

$$PG/CG = w_3/(w_1 + w_2).$$

There is thus a definite point G, fixed with respect to the particles, through which the resultant weight of the particles always passes. It does not depend on the directions of the weights (provided that they are parallel), and thus on the orientation of the body. This point is called the *centre of gravity* of the particles, and it is clear that the argument can be extended to show that such a point exists for any number of particles.

The centre of gravity of a body or a set of particles will be denoted in diagrams by G.

Centre of gravity of a system of coplanar particles

Consider a set of coplanar particles with masses $m_1, m_2, \ldots$ and positions relative to a pair of x and y axes of $(x_1, y_1), (x_2, y_2)$, etc. Let the total mass be M, and the centre of gravity be the point $G(\bar{x}, \bar{y})$ (Fig. 14.2).

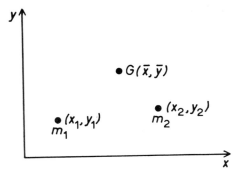

Fig. 14.2

To obtain expressions for $\bar{x}$ and $\bar{y}$ we use the principle

moment of resultant = moment of original system.

Suppose the weights act at right angles to the plane of the paper, so that they have moments about both axes. Applying the principle to the y-axis, we have

$$(m_1 + m_2 + \ldots)g\bar{x} = m_1 gx_1 + m_2 gx_2 + \ldots$$
$$\therefore \; M\bar{x} = \sum mx \qquad \text{(cancelling the } g\text{'s)}$$

and hence

$$\bar{x} = \frac{\sum mx}{M}.$$

Similarly, taking moments about the x-axis, we obtain

$$\bar{y} = \frac{\sum my}{M}.$$

Centre of mass, Centroid

The centre of mass of a system of coplanar particles is *defined* by the above standard equations for $\bar{x}$ and $\bar{y}$. Since the g's have cancelled out these equations make no direct reference to weight, and they are meaningful whether or not the system or body is near a large planet and has weight, and whether or not the weights of the constituent particles are all parallel. The term 'centroid' denotes a *geometric centre*, and it is defined to be coincident with the centre of gravity of a body provided that the matter in that body is uniformly distributed. In practice, for uniform bodies of ordinary size near the surface of the earth, the three points are coincident.

Example   Particles of mass 2 kg, 5 kg and 7 kg are placed at the vertices of an equilateral triangles of side a. Locate their centre of gravity.

If x and y axes are introduced – which is not essential – they may be conveniently placed as shown in Fig. 14.3. The standard equations for $\bar{x}$ and $\bar{y}$ can now be used, but it is probably just as easy here to proceed directly by taking moments about the two axes.

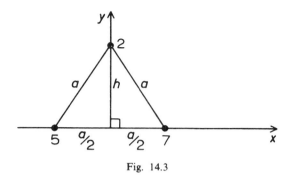

Fig. 14.3

Taking moments about the y-axis we have

$$\frac{7a}{2} - \frac{5a}{2} = 14\bar{x}$$

from which

$$\bar{x} = a/14.$$

Also, since $h = a\sin 60° = a\sqrt{3}/2$, we have, taking moments about the x-axis,

$$2a\sqrt{3}/2 = 14\bar{y},$$

from which

$$\bar{y} = a\sqrt{3}/14.$$

Exercise 14a

General point: If a system of particles is represented as being in the plane of the paper, it is essential in determining their centre of gravity to take moments about axes which are also in the plane of the paper.

1 Particles of mass 3, 7, 4 and 6 kg are placed at the vertices A, B, C, D of a rectangle $ABCD$ in which $AB = 3$ m and $BC = 2$ m. Find the distances of the centre of gravity from AB and BC.

2 Particles of mass $2m$, $3m$, $5m$, $10m$ are placed at the points $(1, 2)$, $(3, -1)$, $(-2, 0)$, $(0, -3)$. Find $\bar{x}$ and $\bar{y}$.

3 Particles of mass $5m$, $10m$, $8m$, $7m$ are placed at the points A, B, C, D of a straight line $ABCD$ in which $AB = BC = CD = a$. Find the distance of the centre of gravity from B.

4 ABC is an isosceles triangle in which $AB = AC = 20$ cm, and $BC = 24$ cm. AD is the perpendicular from A to BC. Particles of mass 6, 3, 2, 4 kg are placed at A, B, C, D. Find the distances of the centre of gravity from AD and BC.

5 Particles of mass 1, 2, 3, 4 kg are placed at the vertices A, B, C, D of a rectangle $ABCD$ in which $AB = 8$ m and $BC = 10$ m. Find the distances of the centre of gravity from AB and BC, and deduce its distance from B.

6 Particles of mass 10, 15, 25 kg are placed at the vertices of an equilateral triangle of side 4 m. Find the distance of the centre of gravity from the 10 kg particle.

7 Particles of mass 2, 4, 3, 4, 2 and 5 kg are placed, in that order, at the vertices of a regular hexagon. Prove that the centre of gravity is at the centre of the hexagon.

8 Particles of mass 3, 4 and m kg are placed at the points A, B, C of a straight line ABC in which $AB = BC = a$. Find m if the centre of gravity is distant $3a/2$ from A.

9 Particles of mass 2 kg and 5 kg are placed at points A and B, 2 m apart. How far from A must a particle of mass 3 kg be placed to give a centre of gravity which is 2.5 m from A on AB produced?

10 Particles of mass 15 and 25 are placed at points $(2, 1)$ and $(3, -2)$, and a particle of mass 10 is placed at the point (h, k). Find h and k if the centre of gravity is at the point $(4, 0)$.

Centres of gravity of some simple 2-dimensional bodies

Certain bodies have clear geometrical centres, and it is apparent that if the matter in them is uniformly distributed, the centres of gravity of such bodies must lie at these points.

Examples are the *uniform rod*, and uniform laminas in the shapes of *circles* and *rectangles*.

The uniform triangular lamina

Consider the lamina to be made up of many narrow strips parallel to BC, as shown in Fig. 14.4. The centre of gravity of each strip is at its mid-point, and the lamina is thus replaceable by a series of particles on the *median AD*. Hence the centre of gravity of the whole lamina lies on AD, and it follows by similar reasoning that it also lies on the other two medians. Combining this conclusion with a well-known geometrical theorem regarding the point of intersection of the medians of any triangle, we have the following result.

The centre of gravity of a uniform triangular lamina lies at the intersection of its medians, i.e. at the point which divides each median in the ratio 1 : 2.

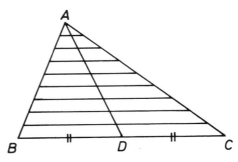

Fig. 14.4

Note: It is easily shown that three equal particles placed at the vertices of a triangle have the same centre of gravity as the triangular lamina.

A uniform lamina in the shape of a *parallelogram* has its centre of gravity at the intersection of its diagonals. This can be proved by a method similar to that used for the triangle.

Compound bodies, remainders

We now consider methods for locating the centres of gravity of bodies which can be formed by adding and subtracting the standard bodies. Bodies of these kinds are known respectively as compound bodies and remainders.

Examples Figure 14.5 shows a compound body (rectangle + triangle), and Fig. 14.6 shows a remainder (rectangle − triangle).

Fig. 14.5

Fig. 14.6

To find the centre of gravity of either kind of body we first suppose each constituent body to be replaced by a particle at its own centre of gravity. The weight of such a particle is proportional to the area of the figure it replaces, and it is in fact usually convenient to use this area as a direct measurement of the particle's weight. Next we use the principle

moment of whole = sum of moments of parts.

In the case of a remainder, this becomes

moment of whole = moment of part removed + moment of remainder.

It should be noted once more that if laminas are represented as lying in the plane of the paper, moments should usually be taken about axes which also lie in that plane.

The method will now be illustrated by worked examples. For simplicity these are limited to 2-dimensional cases (laminas), though it will be clear that the method can also be easily applied to simple solid bodies such as cuboids and spheres.

Example 1 A uniform lamina *ABCED* consists of an isosceles triangle *CED* of a height 15 cm added onto a square *ABCD* of side 10 cm (Fig. 14.7). Locate the centre of gravity.

The dotted line is an axis of symmetry, on which the centre of gravity clearly must lie. To find its position on this line we take moments about *any axis perpendicular to the line*, say *AB*. In order to do this it is convenient to tabulate the required areas and distances, as in Table 14.1.

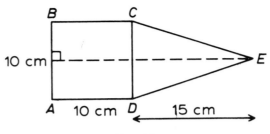

Fig. 14.7

Table 14.1

Figure	Area	Distance of centre of gravity from *AB*
ABCD	100	5
CED	75	$10 + \frac{15}{3} = 15$
ABCED	175	*x*

Now using the principle

moment of whole = sum of moments of parts,

we have, taking moments about *AB*,

$$175x = (100 \times 5) + (75 \times 15),$$

from which

$$x = 9.286 \text{ cm}.$$

Example 2   *ABCD* is a uniform rectangular lamina in which $AB = 30$ cm and $BC = 24$ cm, and *E* is the mid-point of *BC*. The triangle *ABE* is removed (Fig. 14.8). Find the distances of the centre of gravity of the remainder from *AB* and *AD*.

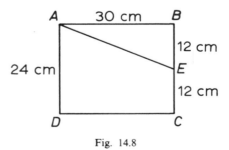

Fig. 14.8

Here it is important to note that since the centre of gravity of triangle *ABE* divides the median through *A* in the ratio $1:2$, its *perpendicular* distance from *BE* is one-third of 30, i.e. 10. Similarly, its perpendicular distance from *AB* is 4. We can thus construct Table 14.2.

Table 14.2

Figure	Area	Distance of centre of gravity from AB	Distance of centre of gravity from AD
$ABCD$	720	12	15
ABE	180	4	20
$AECD$	540	x	y

Now using the principle

moment of whole = moment of part removed + moment of remainder,

we have, taking moments about AB,

$$720 \times 12 = (180 \times 4) + 540x$$

from which

$$x = 14\tfrac{2}{3}\,\text{cm}.$$

Also, taking moments about AD, we obtain

$$720 \times 15 = (180 \times 20) + 540y,$$

from which

$$y = 13\tfrac{1}{3}\,\text{cm}.$$

Example 3 ABC is a uniform triangular lamina in which D and E are the mid-points of AB and AC (Fig. 14.9). Find the position of the centre of gravity of the quadrilateral $BCED$.

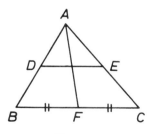

Fig. 14.9

Since the centres of gravity of the triangles ABC and ADE lie on the median AF, that of the quadrilateral also lies on this line. We therefore take moments about any axis which is perpendicular to the line AF, say the axis through A which is perpendicular to the paper. If we choose this axis, we are able to speak simply of 'taking moments about A'.

Triangles ABC, ADE are clearly similar, and since the ratio of their sides is 2 to 1, that of their areas is 4 to 1. It is therefore convenient to let the actual areas be 4 units and 1 unit, respectively. Letting $AF = d$, we ean then tabulate the required data as in Table 14.3.

Table 14.3

Figure	Area	Distance of centre of gravity from A
ABC	4	$2d/3$
ADE	1	$2/3 \times d/2 = d/3$
$BCEF$	3	x

Taking moments about A, we have

$$4 \times 2d/3 = (1 \times d/3) + 3x,$$

from which

$$x = 7d/9.$$

Example 4   Figure 14.10 shows an L-shaped lamina formed by joining the two uniform similar rectangles $ABCD$ and $CEFH$. Given that the density of the first rectangle is twice that of the second, find the position of the centre of gravity.

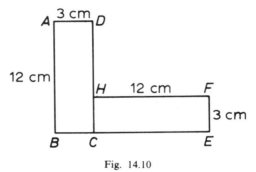

Fig. 14.10

It is convenient here to let the weights of the rectangles be 2 units and 1 unit, and to locate the centre of gravity by obtaining its distances from AB and BE. Let these be x and y.

Taking moments about AB, we have

$$(2 \times 1.5) + (1 \times 9) = 3x,$$

from which

$$x = 4 \text{ cm}.$$

Taking moments about BE, we have

$$(2 \times 6) + (1 \times 1.5) = 3y$$

from which

$$y = 4.5 \text{ cm}.$$

Suspension problems

Problems on the determination of centres of gravity are often followed by ones involving the suspension of the object whose centre of gravity has just been found. In such cases we can immediately put a vertical line into the diagram by joining the point of suspension to the centre of gravity. It is not always necessary to re-orientate the diagram to achieve a physically realistic effect.

Example   If the L-shaped lamina of the last example is suspended from A, what will be the inclination of AB to the vertical?

The simplified diagram of Fig. 14.11 is all that is needed.

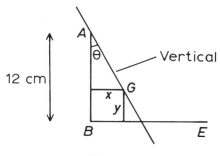

Fig. 14.11

We have

$$\tan \theta = \frac{x}{12-y} = \frac{4}{7.5}$$

Hence

$$\theta = 28.07°.$$

Another popular type of problem which often follows the location of a centre of gravity involves the conditions for a body to *topple*. For worked examples on this topic see Chapter 9.

Exercise 14b

General point: Always look for axes of symmetry; if one exists, take moments about an axis perpendicular to it.

1   A uniform lamina $ABCED$ consists of an isosceles triangle CED attached to a square $ABCD$. The side $AB = 8$ cm and $CE = 5$ cm. Find the distance of the centre of gravity from AB.

2   From a uniform square lamina $ABCD$ of side 12 cm an isosceles triangle ABE of height 9 cm is removed. Find the distance of the centre of gravity of the remainder from CD.

3 $ABCD$ is a uniform square lamina of side 20 cm and E is the mid-point of CD. The triangle ADE is removed. Find the distances of the centre of gravity of the remainder from AB and BC. If the remainder is suspended from B, what will be the inclination of BC to the vertical?

4 $ABCD$ is a uniform rectangular lamina of mass 2 kg in which $AB = 20$ cm and $BC = 12$ cm. Particles of mass 1, 3, 2, 4 kg are placed at A, B, C, D. Find the distances of the centre of gravity from AB and BC.

5 $ABCD$ is a uniform square lamina and E, F are the mid-points of AB, BC. The square $EBFG$ is removed and the remainder is suspended from A. Find the inclination of AD to the vertical.

6 $ABCED$ is a lamina consisting of an isosceles triangle CED attached to a square $ABCD$. The side $AB = 10$ cm and $CE = 13$ cm, and the density of the triangle is twice that of the square. Find (a) the distance of the centre of gravity from AB, (b) the inclination of AB to the vertical when the lamina is suspended from A.

7 $ABCD$ is a uniform rectangular lamina of mass 8 kg in which E is the mid-point of CD. The triangle ADE is cut away and the remainder is placed in a vertical plane with CE on horizontal ground. Find the mass of the greatest particle which can be attached at A without causing the lamina to topple.

8 $ABCD$ is a uniform square lamina of side 20 cm. A circle of radius 7 cm, with its centre 8 cm from AB on the diagonal AC, is removed. Find the distance of the centre of gravity of the remainder from C. (Take π to be 22/7.)

9 ABC is an L-shaped piece of uniform wire of mass 200 g in which $\angle ABC = 90°$, $AB = 60$ cm and $BC = 40$ cm. When A is smoothly hinged to a fixed support, find (a) the angle between AB and the vertical when it is hanging freely, (b) the minimum force which must be applied at C to make the wire hang with AB vertical.

10 ABC is a uniform triangular lamina in which the length of the perpendicular from A to BC is h. The points D and E are on AB and AC respectively, such that DE is parallel to BC and $AD = AB/4$. Find in terms of h the distance of the centre of gravity of $BCED$ from BC.

11 $ABCD$ is a uniform rectangular lamina in which $AB = 1\frac{1}{2}BC$. A triangular lamina of three times its density is superimposed upon it, just covering ABD. If the compound body is suspended from A, what will be the inclination of AB to the vertical?

12 $ABCD$ is a uniform rectangular lamina of mass 2 g in which $AB = 20$ cm and $BC = 30$ cm. Wire of density 10 g m^{-1} is attached along the sides AB, BC, CD. If A is smoothly hinged to a fixed support, find (a) the inclination of AB to the vertical when the lamina is hanging freely, (b) the vertical force at D which is needed to make it hang with AD horizontal.

13 $ABCDEF$ is a uniform regular hexagonal lamina of side a. If the triangle ABC is removed find the distance of the centre of gravity of the remainder from E.

14 ABC is a uniform triangular lamina in which $AB = AC$ and D is the foot of the perpendicular from A to BC. It is suspended by means of vertical strings attached at A and B, and hangs with AB horizontal. Prove that if the ratio of the tensions in the strings is 2 to 3, the ratio of AD to BC is 3 to 2.

15 E is the centre of a uniform square lamina $ABCD$ of side $6a$ and mass M. The triangle BEC is removed and superimposed on AED. Find (a) the distance of the centre of gravity from AD, (b) the minimum force at B which makes the body topple about D when it is standing in a vertical plane with CD on horizontal ground.

16 $ABCD$ is a uniform rectangular lamina in which $AB = 20$ cm and $BC = 7$ cm. The circle touching the sides AB, BC and CD is removed and superimposed on the lamina so as to touch the sides AB, AD and CD. Find the distance of the centre of gravity from AD. (Take π to be 22/7.)

17   Two identical rods, AB and BC, are joined together at B, and a particle of the same mass is attached at A. Find the maximum value of the angle ABC if toppling does not occur when the rods are placed in a vertical plane with BC in contact with horizontal ground.

18   A uniform lamina is in the form of a trapezium $ABCD$ in which $AB = BC = CD = 10$ cm, and the parallel sides AD, BC are 8 cm apart. The lamina is placed in a vertical plane with BC resting on a surface which is steadily tilted. Assuming that the lamina does not slide, at what inclination to the horizontal will it topple?

19   A uniform square lamina $ABCD$ of side a stands in a vertical plane with AB on horizontal ground. P and Q are points on the sides AB, AD such that $PB = QD = x$. Show that if the square $APRQ$ is removed, the lamina will topple unless $a^2 - 3ax + x^2 \leqslant 0$.

20   An open container consists of a uniform hollow cylinder of radius R fixed to a uniform circular base of the same radius and $1\frac{1}{2}$ times the density. Prove that if the container is on the point of toppling when standing on a plane inclined at arc tan 2 to the horizontal, then its height is $3R/2$.

The use of calculus to find centres of gravity

The equation $\bar{x} = \Sigma\, mx/M$ was obtained above for a system of *particles*. To find the centres of gravity of bodies consisting of continuously distributed matter we need in general to divide the bodies into infinitesimally small elements. In the expression $\Sigma\, mx$, therefore, the m's tend to zero, the number of terms tends to infinity, and the whole expression thus becomes an *integral* of the form $\int x\, dm$. Such integrals are evaluated by ordinary calculus techniques. The method will be illustrated by applying it to a few simple standard bodies.

Uniform circular arc

Consider an arc of a circle of radius a which subtends an angle of 2α at the centre O (Fig. 14.12). Since the arc is uniform we can use length as a measure of mass. By

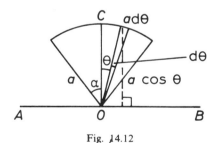

Fig. 14.12

symmetry the centre of gravity lies on OC. Dividing the arc into infinitesimal elements as shown, we have

$$\text{mass of element} = a\, d\theta$$

$$\therefore \text{ moment of element about } AB = a\, d\theta\, a\cos\theta$$

$$= a^2 \cos\theta\, d\theta.$$

$$\therefore \text{ total moment} = \int_{-\alpha}^{\alpha} a^2 \cos\theta\, d\theta$$

$$= [a^2 \sin\theta]_{-\alpha}^{\alpha}$$

$$= 2a^2 \sin\alpha.$$

Now total mass = total length = $2a\alpha$. Hence

$$\text{distance of centre of gravity from } AB = \frac{\text{total moment}}{\text{total mass}}$$

$$= \frac{a \sin \alpha}{\alpha}.$$

Uniform sector

The procedure here is to divide the sector into infinitesimally narrow arcs (Fig. 14.13) and use the result just obtained.

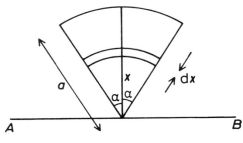

Fig. 14.13

The centre of gravity again lies on the axis of symmetry. Using area as a measure of mass we have

$$\text{mass of elementary arc} = 2\alpha x \, dx$$

$$\therefore \quad \text{moment of arc about } AB = 2\alpha x \, dx \frac{x \sin \alpha}{\alpha} \quad \text{(using our last result)}$$

$$= (2 \sin \alpha) x^2 \, dx.$$

$$\therefore \quad \text{total moment about } AB = 2 \sin \alpha \int_0^a x^2 \, dx = \frac{2a^3 \sin \alpha}{3}.$$

Also

$$\text{total mass of sector} = \frac{a^2 \times 2\alpha}{2} \quad \text{(using the formula } r^2\theta/2 \text{ for the area of a sector)}$$

$$= a^2 \alpha.$$

Hence
distance of centre of gravity from AB = total moment/total mass

$$= \frac{2a \sin \alpha}{3\alpha}.$$

Note: This result can also be obtained by dividing the sector into elementary sectors, treating each sector as a triangle and using the result for the centre of gravity of a triangle.

Uniform solid right cone

Let the cone have a height of H and a base-radius of R. By symmetry the centre of gravity is on the axis of the cone, so we divide it into elementary discs which are perpendicular to this axis, as shown in Fig. 14.14.

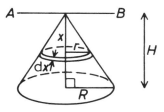

Fig. 14.14

Using volume as a measure of mass, we have

$$\text{mass of elementary disc} = \pi r^2 \, dx$$

$$\therefore \text{ moment of disc about } AB = \pi r^2 x \, dx.$$

Now by similar triangles

$$\frac{r}{R} = \frac{x}{H}$$

$$\therefore \text{ moment} = \pi R^2 x^3 / H^2$$

$$\therefore \text{ total moment} = \frac{\pi R^2}{H^2} \int_0^H x^3 \, dx$$

$$= \frac{\pi R^2}{H^2} \cdot \frac{H^4}{4} = \frac{\pi R^2 H^2}{4}.$$

Also the total mass of the cone is $\frac{1}{3}\pi R^2 H$. Hence

distance of centre of gravity from vertex of cone = total moment/total mass

$$= \frac{3H}{4}.$$

Uniform solid hemisphere

Let the hemisphere have a radius of a. Again there is an axis of symmetry and we divide the hemisphere into elementary discs which are perpendicular to this axis (Fig. 14.15).

Using volume as a measure of mass, we have

$$\text{mass of elementary disc} = \pi r^2 \, dx$$

$$= \pi(a^2 - x^2) \, dx.$$

Hence

$$\text{moment of disc about } AB = \pi x(a^2 - x^2) \, dx$$

$$= \pi(a^2 x - x^3) \, dx.$$

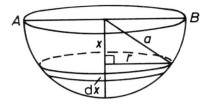

Fig. 14.15

Hence

$$\text{total moment} = \pi \int_0^a (a^2 x - x^3)\, dx$$

$$= \pi \left[a^2 \frac{x^2}{2} - \frac{x^4}{4} \right]_0^a$$

$$= \pi a^4 / 4.$$

Also

$$\text{total mass of hemisphere} = 2\pi a^3 / 3.$$

$\therefore$ distance of centre of gravity from AB = total moment/total mass

$$= 3a/8.$$

Centroids of areas defined by graphs and volumes of revolution

Note: The term 'centroid' is more appropriate than 'centre of gravity' for these cases, at least when the figure referred to does not denote a real physical object. Since centroids are defined purely *geometrically*, we can determine them by taking moments of areas and volumes rather than weights and masses.

The method will be illustrated by worked examples.

Example 1   Find the centroid of the area between the graph of $y = x^2$, the x-axis, and the line $x = 2$ (Fig. 14.16).

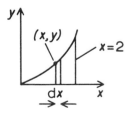

Fig. 14.16

Let the centroid be the point $(\bar{x}, \bar{y})$. Dividing the area into elementary strips parallel to the y-axis, as shown, we proceed as follows.

To find $\bar{x}$

$$\text{Area of elementary strip} = y\,dx$$
$$\therefore \text{ moment of strip about } y\text{-axis} = xy\,dx$$
$$\therefore \text{ total moment about } y\text{-axis} = \int_0^2 xy\,dx$$
$$= \int_0^2 x^3\,dx$$
$$= [x^4/4]_0^2$$
$$= 4.$$
$$\text{Also total area} = \int_0^2 y\,dx$$
$$= \int_0^2 x^2\,dx$$
$$= [x^3/3]_0^2 = \tfrac{8}{3}.$$

We thus have

$$\bar{x} = \frac{\text{total moment about } y\text{-axis}}{\text{total area}} = 4 \times \tfrac{3}{8} = \mathbf{1.5}.$$

To find $\bar{y}$

$$\text{Area of elementary strip} = y\,dx$$
$$\text{Distance of centroid of strip from } x\text{-axis} = y/2$$
$$\therefore \text{ moment of strip about } x\text{-axis} = \frac{y^2\,dx}{2}$$
$$\therefore \text{ total moment about } x\text{-axis} = \int_0^2 \frac{y^2\,dx}{2}$$
$$= \int_0^2 \frac{x^4\,dx}{2}$$
$$= \left[\frac{x^5}{10}\right]_0^2$$
$$= \tfrac{32}{10} = \tfrac{16}{5}$$

We now have

$$\bar{y} = \frac{\text{total moment about } x\text{-axis}}{\text{total area}} = \tfrac{16}{5} \times \tfrac{3}{8} = \mathbf{1.2}.$$

Points to note regarding centroids of areas

1. Sometimes a decision has to be made whether to divide the area into vertical or horizontal strips. Consider for example the area between the graph of $y = x^2$, the y-axis and the line $y = 3$ (Figs. 14.17, 14.18).

In Fig. 14.17 each strip has an area of $x\,dy$, and if we choose this method we must express x as a function of y before integrating and integrate between y limits.

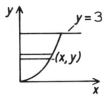

Fig. 14.17

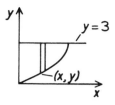

Fig. 14.18

In Fig. 14.18 each strip has an area of $(3 - y)\,dx$, and if this method is adopted y must be expressed as a function of x, as it already is in the equation of the curve.

Which method is preferable depends in general on the region whose centroid is required and on the equation of the curve.

2. Many areas have axes of symmetry, giving $\bar{x}$ or $\bar{y}$ by inspection. For example, the area between the x-axis and the parabola $y = (x - 1)(3 - x)$ has an axis of symmetry $x = 2$ (Fig. 14.19). Hence $\bar{x} = 2$.

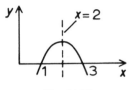

Fig. 14.19

Example 2 The graph of $y^2 = 6x$ is completely rotated about the x-axis. Find the centroid of the solid region between $x = 0$ and $x = 2$ (Fig. 14.20).

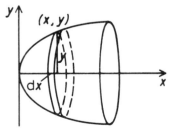

Fig. 14.20

By symmetry $\bar{y} = 0$, as will clearly be the case for all volumes of revolution about the x-axis. To find $\bar{x}$ we divide the volume into vertical elementary discs as shown and proceed as follows.

$$\text{Volume of elementary disc} = \pi y^2\,dx$$
$$= 6\pi x\,dx.$$
$$\therefore\ \text{moment of disc about } y\text{-axis} = 6\pi x^2\,dx.$$

$$\therefore \text{ total moment about } y\text{-axis} = \int_0^2 6\pi x^2 \, dx$$
$$= [2\pi x^3]_0^2$$
$$= 16\pi.$$

Also

$$\text{total volume} = \int \pi y^2 \, dx$$
$$= \int_0^2 6\pi x \, dx$$
$$= [3\pi x^2]_0^2$$
$$= 12\pi.$$

Hence

$$\bar{x} = \frac{\text{total moment about } y\text{-axis}}{\text{total volume}} = \frac{16\pi}{12\pi} = 1\tfrac{1}{3}.$$

Before the final exercise it will be convenient to tabulate (Table 14.4) the centres of gravity of some standard uniform bodies. Most but not all of these results are proved above; all are quotable without proof in problems unless the proof is specifically asked for.

Table 14.4

Body	Position of centre of gravity
Triangle	At intersection of medians
Parallelogram	An intersection of diagonals
Arc	$\dfrac{a \sin \alpha}{\alpha}$ from centre
Sector	$\dfrac{2a \sin \alpha}{3\alpha}$ from centre
Cone, Pyramid	$H/4$ from base
Solid hemisphere	$3a/8$ from centre
Hollow hemisphere	$a/2$ from centre

Exercise 14c

In questions 1–7, find the centroids of the areas bounded by the given graphs and lines.

1 $y = x^3$, the x-axis, and the line $x = 1$.
2 $y = x^2$, the x-axis, and the lines $x = 1$, $x = 2$.
3 $y = x(x - 2)$ and the x-axis.
4 $y = x^2$, the y-axis, and the line $y = 4$.
5 $y^2 = 4x$, the y-axis, and the line $y = 2$.
6 $y^2 = 1 - x$ and the y-axis.
7 $y = e^x$, the x and y axes, and the line $x = 1$. (Leave the answer in terms of e.)

8   Prove by integration that the centre of gravity of a uniform semicircular arc of radius a is $2a/\pi$ from the centre of the circle of which it is a part.

9   Prove by integration that the centre of gravity of a uniform semicircular lamina of radius a is $4a/3\pi$ from the centre of the circle of which it is a part.

10   Wire is mounted along the curved part of the circumference of a uniform semicircular lamina. The mass of the wire is 2 g and that of the lamina is 6 g. If the lamina is suspended from one of the ends of its straight edge, what will be the inclination of this edge to the vertical?

11   A uniform rectangular lamina of width d is attached to a uniform semicircular lamina of the same density and diameter d, forming a compound lamina in which the two edges of length d are coincident. Find the length of the rectangle if the centre of gravity is on the common edge.

In questions 12–16, find the x values of the centroids of the solids of revolution formed by rotating about the x-axis the regions bounded by the given graphs and lines.

12 $y^2 = 4x$ and the line $x = 1$.

13 $y = 2x$, the x-axis, and the lines $x = 1$, $x = 2$.

14 $y = x^2$, the x-axis, and the line $x = 2$.

15 $y^2 = 2x^3$, the x-axis, and the line $x = 1$.

16 $y = e^x$, the x and y axes, and the line $x = 1$.

17   A uniform solid hemisphere of radius R is attached to a uniform right solid cone of equal density with radius R and height $2R$, the two plane faces being coincident. If the compound body is suspended from a point on the rim of the common face, what will be the inclination of this face to the vertical?

18   From a uniform solid cylinder of radius a and height $2a$, a solid hemisphere of the same radius is removed, the plane face of the hemisphere being one of the original plane faces of the cylinder. The remaining solid is placed with its circular base on a horizontal plane, and this plane is steadily tilted. At what inclination will the body topple if it does not first slide?

19   The circle with equation $x^2 + y^2 = a^2$ is rotated about the x-axis, forming a uniform solid sphere of radius a. By integrating between the limits $a/2$ and a, obtain the value of $\bar{x}$ for the cap of the sphere of depth $a/2$, and deduce the distance of the centre of gravity of the cap from its plane face.

20   A toy consists of a uniform solid hemisphere of radius R, attached to a solid right uniform cone of base radius R and twice the density. The two circular plane faces coincide. When the toy is placed with any part of its spherical surface in contact with horizontal ground, it stands in equilibrium. Find the height of the cone in terms of R.

21   The area between the graph of $y = 2x$, the x-axis, and the lines $x = 1$ and $x = 3$ is rotated about the x-axis, forming a solid uniform frustum of a cone. If such a frustum is placed with its smaller circular face in contact with a horizontal plane, and the plane is steadily tilted, at what inclination will the frustum topple if it does not first slide?

22   A uniform solid right cone of semi-vertical angle arc $\tan \frac{1}{2}$ is suspended by two vertical strings, one attached to its vertex and one to a point on its circular rim. Find the inclination of the plane face to the vertical if the ratio of the tensions is $1:2$.

23   A uniform hollow hemisphere of radius R is joined to a uniform solid right cone of base radius R and the same mass, the circular edge of the hemisphere being coincident with that of the cone. The body is suspended from a point on the common edge, and hangs with the lower slant edge of the cone horizontal. Prove that the height of the cone is $4R$.

24 The cap of question 19 is mounted on a uniform cylinder of equal density whose cross section is equal to that of the cap's plane face. Prove that if the centre of gravity of the compound body lies on the common circular face, then the length of the cylinder is $\frac{1}{6}a\sqrt{3.5}$.

25 Across a diameter of the circular rim of a uniform hollow hemispherical vessel is fixed a uniform rod which is equal in length to the diameter and of half the hemisphere's mass. The vessel is then suspended by two vertical strings, one attached to the rim and one to the centre of the rod. If the ratio of the tensions is $3:2$, what is the angle of inclination of the rod to the vertical?

26 The radii of the end faces of the frustum of a uniform right solid cone are in the ratio $1:2$, and the faces are a distance d apart. Prove that if the frustum is on the point of toppling when placed with its curved surface in contact with horizontal ground, then the radius of the smaller face is $d\sqrt{\frac{17}{28}}$.

15

Probability

Some terminology

Possibility space

Probability theory deals with situations and experiments in which a number of different *outcomes* are possible. *A possibility space is a complete and mutually exclusive set of such outcomes.*

Consider for example a throw of a die. The simplest possibility space is the set

$$\{1, 2, 3, 4, 5, 6\},$$

and some others are

$$\{\text{odd, even}\} \quad \text{and} \quad \{\text{prime, non-prime}\}.$$

The following, however, are not possibility spaces:

$$\{\text{even, prime}\}, \quad \{3, \text{odd, even}\}.$$

The first of these sets is not complete, since '1' is not covered by either outcome, and the second set is not mutually exclusive, since '3' is part of the outcome 'odd'.

A possibility space is denoted by $\mathscr{E}$, the symbol for a universal set.

Event

An event is a happening associated with a subset of a possibility space. For example, the event 'getting less than 3' with a die is associated with the subset $\{1, 2\}$ of the possibility space $\{1, 2, 3, 4, 5, 6\}$. Events are denoted by capital letters, and we shall use the same capital letter to denote an event and the subset associated with it. Thus we might have both

$$A = \text{the event 'getting less than 3'}$$
$$\text{and} \quad A = \text{the subset } \{1, 2\}.$$

The probability of an event

To define the probability of an event we first need a possibility space in which *all the outcomes are equally likely* ('equiprobable'). Given this, the probability of an event A, denoted by $p(A)$, is defined as follows.

$$p(A) = \frac{\text{the number of equiprobable outcomes in the subset } A}{\text{the number of equiprobable outcomes in the possibility space}}.$$

More briefly, using the terminology of set theory, we have

$$p(A) = \frac{n(A)}{n(\mathscr{E})}.$$

For example, the probability of getting an odd number with a (fair) die is $\frac{3}{6} = \frac{1}{2}$, and the probability of getting an ace when a card is drawn at random from an ordinary pack is $\frac{4}{52} = \frac{1}{13}$. The values of these fractions clearly give measurements of the likelihoods or probabilities of the respective events. It follows from the definition that all probabilities are between 0 and 1, and that the sum of the probabilities of all the outcomes in a possibility space is 1.

The requirement that all outcomes be equiprobable, for the definition to be applicable, is very important, and neglect of it leads to some of the most common errors in problems on probability. To take a simple example, suppose that a penny is thrown twice and we require the probability of getting both a head and a tail. It is tempting to think that the probability is $\frac{1}{3}$, on the grounds that the outcomes could be described as

two heads,　two tails,　head and tail.

This reasoning is incorrect, however, since these outcomes are not equiprobable. The set of equiprobable outcomes is

HH, TT, HT, TH,

from which it is clear that the probability of a head and a tail is $\frac{2}{4} = \frac{1}{2}$.

An alternative definition of probability

If we threw a fair die a large number of times we should expect any given number, say a 2, to occur on about $\frac{1}{6}$ of the throws, that is the same fraction as the probability of a 2 on a single throw. It can in fact be shown that this fraction, the *relative frequency* of the event 'getting a 2', approaches $\frac{1}{6}$ as a limit as the number of throws tends to infinity. We thus have the following alternative definition of probability.

> **The probability of an event A is the limit to which the relative frequency of A tends as the number of trials tends to infinity.**

When we know that all the outcomes in a possibility space are equally likely on account of symmetry, as in the case of a throw of a fair die or a random draw of a card from a pack, the first definition, $n(A)/n(\mathscr{E})$, gives *exact* values of the probabilities and reliance on the second definition is unnecessary. But if we have a biased die, or if, for example, we are interested in our chance of being struck by lightning, we can only *estimate* the required probability by finding the relative frequency of the event in a large number of trials.

The negation of an event

Given an event A, associated with a subset of a possibility space, we define the *negation* of A, called 'not-A', as the event associated with the *complement* of the set, A'. Since $p(A) + p(A') = 1$, we have

$$\boxed{p(A') = 1 - p(A).}$$

For example, the probability of not getting an ace, on a random draw from an ordinary pack, is $1 - \frac{1}{13} = \frac{12}{13}$. The probability of the negation of an event is often easier to calculate than that of the event itself (see Example 1(b), below).

The event 'A and B'

Suppose that a die and a coin (both 'fair') are thrown together, and A, B are the events 'getting a 4' and 'getting a head', respectively. The compound event 'A and B' is then the event 'getting *both* a 4 and a head', and we can obtain its probability by the following reasoning.

If a die and a coin are thrown a large number of times, a 4 will occur on $\frac{1}{6}$ of the throws (approximately). On half of these occasions a head will also occur, so both events will occur on $\frac{1}{6} \times \frac{1}{2} = \frac{1}{12}$ of the throws. It follows that $\frac{1}{12}$ is the probability that both events occur, and since $\frac{1}{6}$ and $\frac{1}{2}$ are the probabilities of the individual events A and B, we have

$$p(A \text{ and } B) = p(A) \times p(B).$$

This is an important probability theorem, but it does not hold in all cases. It is valid only when A and B are *independent* events, that is when the occurrence of one of them does not affect the probability of the other. To see how the reasoning must be modified when the events are not independent, we will take another example. Suppose that six equally matched children play in a competition, and we require the probability that John wins (A) and Jane comes second (B).

It is important to see first that in this case A and B are not independent events. If nothing is known about the result of the competition, then Jane is as likely to be second as to occupy any other place, so the probability that she is second is $\frac{1}{6}$. But if it is known that John has won, that is that event A has definitely occurred, then there are only five children who can be second, so the probability that Jane is second is $\frac{1}{5}$. We can express this symbolically as follows.

$$p(B) = \frac{1}{6}, \quad p(B|A) = \frac{1}{5}.$$

The symbol $p(B|A)$ denotes *the probability of B, given that A is known to have occurred*, and it is called a *conditional* probability. Consider now the probability of the compound event A and B. If a large number of competitions are conducted, John will win $\frac{1}{6}$ of them, and Jane will be second in $\frac{1}{5}$ of the subset in which John wins. Hence both events occur in $\frac{1}{6} \times \frac{1}{5} = \frac{1}{30}$ of the competitions, and it follows that

$$p(A \text{ and } B) = p(A) \times p(B|A).$$

Summing up, we have the following two standard results.

> **If A and B are independent events,**
> **$p(A \text{ and } B) = p(A) \times p(B).$**

> **If A and B are not independent events,**
> **$p(A \text{ and } B) = p(A) \times p(B|A).$**

The event '*A* and *B*' in set terminology

Consider once more the experiment in which a die and a coin are thrown together, and let A, B be the events 'getting a 4' and 'getting a head', respectively. There are 12 outcomes in the complete possibility space, which could be represented as follows:

$$\{1H, 1T, 2H, 2T, 3H, 3T, 4H, 4T, 5H, 5T, 6H, 6T\}.$$

The events A and B are represented by the subsets

$$\{4H, 4T\}, \quad \{1H, 2H, 3H, 4H, 5H, 6H\},$$

which intersect or overlap as illustrated by the following Venn diagram (Fig. 15.1). Clearly the event A and B, which in this case consists of the single outcome 4H, is represented by the *intersection* of the two sets, $A \cap B$.

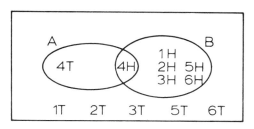

Fig. 15.1

Example 1 A coin is thrown 4 times. Find the probability that (a) 4 heads are obtained, (b) at least one head is obtained.

(a) The outcomes of each throw are clearly independent; hence we have

$$p(4 \text{ heads}) = \frac{1}{2} \times \frac{1}{2} \times \frac{1}{2} \times \frac{1}{2} = \left(\frac{1}{2}\right)^4 = \frac{1}{16}.$$

(b) When the term 'at least' is used in describing an event, it is usually best to calculate the probability of the *negation* of the event. 'At least one head' is a complicated event because it includes the cases in which 1 head, 2 heads, 3 heads and 4 heads occur. The negation, however, is the simple event 'all tails', whose probability is $\frac{1}{16}$ by the reasoning given in (a). Hence

$$p\,(\text{at least one head}) = 1 - \frac{1}{16} = \frac{15}{16}.$$

Example 2   Three cards are drawn at random from an ordinary pack. Find the probability that all are hearts.

In this case the three events are not independent. The probability that the first card is a heart is clearly $\frac{1}{4}$. Given that this occurs, there are 12 hearts left and 51 cards altogether, so the probability that the second card is a heart is $\frac{12}{51}$. Given that both of the first two cards are hearts, the probability that the third card is a heart is $\frac{11}{50}$. It follows that the required probability is

$$\frac{1}{4} \times \frac{12}{51} \times \frac{11}{50} = \frac{11}{850}.$$

The theory of *permutations and combinations* is sometimes useful in problems on probability. This is illustrated in the next example.

Example 3   Three cards are drawn at random from a reduced pack consisting of the aces, kings and queens of each suit. Find the probability that exactly two of them are aces.

It is possible to solve this problem by using probability theorems, but this is somewhat complicated as there are three different ways in which the required hand can be obtained, namely

ace, ace, non-ace;   ace, non-ace, ace;   non-ace, ace, ace.

In the permutations and combinations method, on the other hand, we ignore the order of the cards and simply calculate the total number of hands of the required kind. Since 2 aces have to be selected from 4, and 1 non-ace from 8, the number of hands is $_4C_2 \times 8$. There are $_{12}C_3$ possible 3-card hands altogether, so the probability is

$$\frac{_4C_2 \times 8}{_{12}C_3} = \frac{6 \times 8}{220} = \frac{12}{55}.$$

Exercise 15a

(In this and subsequent exercises, assume that coins and dice are properly constructed.)

1   A card is drawn at random from an ordinary pack. Find the probability that it is (a) a black ace, (b) not an ace or a king, (c) a diamond or a jack or both, (d) black or an ace, but not both.

2   Three coins are thrown. List the possibility space which contains all the possible outcomes, and hence find the probability of (a) exactly two heads, (b) not more than one head.

3   Two dice are thrown. Find the probability of (a) a 6 followed by an odd number, (b) a total of 8, (c) a total of 10 or more, (d) a total of 4 or more.

4 A coin is thrown repeatedly. Find the probability that the first head occurs (a) on the fourth throw, (b) at some time after the second throw.

5 Peter and John both fire at a target, their chances of a hit being $\frac{2}{3}$ and $\frac{3}{4}$, respectively. Find the probability that (a) Peter hits and John misses, (b) both miss, (c) at least one misses.

6 A and B are independent events. (a) Given that $p(A) = \frac{2}{3}$ and $p(B) = \frac{9}{10}$, find $p(A \cap B)$. (b) Given that $p(A') = \frac{4}{5}$ and $p(A \cap B) = \frac{3}{20}$, find $p(B)$.

7 Three balls are drawn at random, without replacement, from a bag containing 4 black and 2 white balls. Find the probability that (a) all are black, (b) the sequence is white, black, white, (c) at least one is white.

8 A die is thrown three times. Find the probability of (a) three numbers of more than 4, (b) at least one number of less than 3.

9 When a biased coin is thrown 6000 times, 4000 tails are obtained. Estimate the probability that in four throws there is at least one head.

10 A committee of three is chosen at random from 6 men and 4 women. Find the probability that (a) all are men, (b) there is at least one man.

11 A 4-card hand is dealt from a reduced pack consisting of the aces, kings and queens of each suit. Use the theory of permutations and combinations to find the probability that it contains (a) a card of each suit, (b) 3 aces and 1 king.

12 I throw three darts at the 'bull', my probability of success on any single throw being x. Write down expressions for the probabilities that I miss (a) three times, (b) at least once.

13 I take a taxi to work on 3 wet days out of every 5, and on 2 dry days out of every 7. If 3 days in every 10 are wet, what is the probability that (a) it is dry and I take a taxi, (b) it is wet and I do not take a taxi?

14 Balls are drawn at random, without replacement, from a bag containing 3 red balls, 2 white balls and 1 blue ball. Find the probability that (a) the blue ball is drawn last, (b) the second ball drawn is white, (c) the second ball drawn is white, given that the first is red, (d) the first three balls drawn are blue, white, red in that order, (e) at least one of the first two balls drawn is white.

15 John, Harry, Jane and Mary are placed at random along one side of a straight table. Use the theory of permutations and combinations to find the probability that (a) Jane and Mary sit together, (b) John and Harry sit at the ends.

16 In a class of 20 boys there are 12 who like football, 11 who like cricket and 2 who like neither of the sports. Draw a Venn diagram and use it to find the probability that a boy chosen at random likes (a) both sports, (b) only football. Find also (c) the probability that of two boys chosen at random, the first likes only cricket and the second likes only football.

17 My chance of beating a certain opponent at chess is $\frac{1}{10}$. Find the probability, to 2 d.p., that I achieve no wins in (a) 3 games, (b) 4 games, (c) 5 games. Find also (d) the number of games I need to play in order that my chance of achieving at least one win is more than 0.5.

18 In a certain country 10% of the men and 20% of the women are left-handed. Out of 10 000 married couples, how many would be expected to have (a) both partners left-handed, (b) at least one partner left-handed?

19 What is the probability that (a) two people, (b) three people, were born on different days of the week?

20 Three numbers are selected at random from the first 10 natural numbers. (No number can be selected more than once.) Find the probability that (a) 2, 5 and

7 are chosen (in any order), (b) all three numbers are odd, (c) two numbers are over 6 and one is under 3.

21   All of the 24 students in a certain class study at least one of the subjects Maths, Physics, Chemistry. Equal numbers study Physics and Chemistry, and all the Physics students study Maths. Given that 4 students study Physics and Chemistry, 6 study Maths and Chemistry and 5 study Chemistry only, draw a Venn diagram and use it to find the probability that a student chosen at random studies (a) Maths but not Chemistry, (b) Physics, Chemistry or both. Find also (c) the probability that both of two students chosen at random study Maths only.

22   Single cards are drawn at random from an ordinary pack, each card being replaced in the pack before the next draw is made. How many draws are needed to make it more likely than not that (a) at least one club will occur, (b) at least one ace will occur?

The event '*A* or *B*'

In probability theory '*A* or *B*' is always taken to mean '*A* or *B* or both', and it is associated with the set $A \cup B$. To illustrate this, consider a throw of a die and let A be the event 'getting an odd number' and B the event 'getting a 2, 3, 5 or 6'. The Venn diagram is shown in Fig. 15.2.

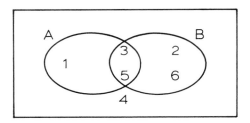

Fig. 15.2

'*A* or *B*' (or both) is clearly the event 'getting a 1, 2, 3, 5 or 6', and the Venn diagram shows that it corresponds to the *union* of the sets A and B. It can also be seen immediately that the probability of A or B is $\frac{5}{6}$, but we require a general method of calculating $p(A \text{ or } B)$, in which this probability is expressed in terms of $p(A)$ and $p(B)$. It is therefore necessary to express $n(A \cup B)$ in terms of $n(A)$ and $n(B)$.

We have $n(A) = 3$ and $n(B) = 4$. To obtain $n(A \cup B)$ we can try adding these together, but if we do so we are counting the elements in the 'overlap' region twice. We therefore obtain $n(A \cup B)$ by *adding* $n(A)$ *and* $n(B)$ *and subtracting* $n(A \cap B)$. This gives us the method of expressing $p(A \text{ or } B)$ in terms of $p(A)$ and $p(B)$. We have

$$p(A \text{ or } B) = \frac{3+4-2}{6}$$

$$= \frac{3}{6} + \frac{4}{6} - \frac{2}{6}$$

$$= p(A) + p(B) - p(A \text{ and } B).$$

This is an important probability theorem. It can be expressed either in words or in set terminology:

$$p(A \text{ or } B) = p(A) + p(B) - p(A \text{ and } B)$$
$$\text{or} \quad p(A \cup B) = p(A) + p(B) - p(A \cap B).$$

An important special case is that in which A and B are *mutually exclusive* events; that is events which cannot occur together. In this case the sets in the Venn diagram do not overlap and we have $p(A \cap B) = 0$. The above theorem thus reduces to the following.

If A and B are mutually exclusive events,
$$p(A \text{ or } B) = p(A) + p(B).$$

For example, if A and B are the events 'getting a 1 or a 2' and 'getting a 4', respectively, with a throw of a die, the probability of A or B is clearly $\frac{2}{6} + \frac{1}{6} = \frac{3}{6} = \frac{1}{2}$, that is $p(A) + p(B)$ (Fig. 15.3).

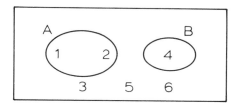

Fig. 15.3

Note: The negation of the event 'A or B' is 'neither A nor B'. As illustrated in the next example, therefore, it is often possible to obtain the probability of an event of this kind by two alternative methods.

Example 4 A card is drawn at random from an ordinary pack, and a coin is thrown. Find the probability of a spade or a head or both.

Method (a)
Letting the two events be A and B, respectively, we have $p(A) = \frac{1}{4}$, $p(B) = \frac{1}{2}$, and since the events are clearly independent, $p(A \text{ and } B) = \frac{1}{8}$. Hence

$$p(A \text{ or } B) = p(A) + p(B) - p(A \text{ and } B)$$

$$= \frac{1}{4} + \frac{1}{2} - \frac{1}{8}$$

$$= \frac{5}{8}.$$

Method (b)
The negation of 'A or B' is 'neither A nor B', that is 'not-A and not-B'. Now $p(\text{not-}A) = 1 - p(A) = \frac{3}{4}$ and $p(\text{not-}B) = 1 - p(B) = \frac{1}{2}$; hence

$$p\,(\text{not-}A \text{ and not-}B) = \frac{3}{4} \times \frac{1}{2} = \frac{3}{8}$$

$$\therefore\ p\,(A \text{ or } B) = 1 - \frac{3}{8} = \frac{5}{8}.$$

It often happens that an event can occur in a number of mutually exclusive ways, and we obtain its probability by adding the probabilities of the component events. The next two examples illustrate this technique.

Example 5 Balls are drawn at random, without replacement, from a bag containing 3 black balls, 2 white balls and 1 green ball. Find the probability that the second ball differs in colour from the first.

This event can occur in three mutually exclusive ways, namely

> A: black, non-black;
> B: white, non-white;
> C: green, non-green.

We have

$$p\,(A) = \frac{3}{6} \times \frac{3}{5} = \frac{3}{10},$$

$$p\,(B) = \frac{2}{6} \times \frac{4}{5} = \frac{4}{15},$$

$$p\,(C) = \frac{1}{6} \times 1 = \frac{1}{6}.$$

Since A, B and C are mutually exclusive events, we obtain the probability of A or B or C by adding the individual probabilities:

$$p\,(A \text{ or } B \text{ or } C) = p\,(A) + p\,(B) + p\,(C)$$

$$= \frac{9}{30} + \frac{8}{30} + \frac{5}{30}$$

$$= \frac{22}{30} = \frac{11}{15}.$$

Example 6 The probability that a given person has blue eyes is x. Obtain an expression for the probability that exactly 3 of a family of 5 children have blue eyes.

This event can occur in many different ways, some of which are shown in the following table, in which B denotes possession of blue eyes.

1st child	2nd	3rd	4th	5th
B	B	B	not-B	not-B
B	B	not-B	B	not-B
B	B	not-B	not-B	B

Since $p(B) = x$ and $p(\text{not-}B) = 1 - x$, the probability of the sequence shown in each row is $x^3(1-x)^2$. The rows clearly represent a set of mutually exclusive outcomes, so we obtain the required probability by adding their individual probabilities together, that is by multiplying $x^3(1-x)^2$ by the total number of rows. The number of rows is the number of ways of making a selection of 3 things from 5 things, namely $_5C_3$ or $_5C_2$. Hence the required probability is

$$_5C_2 x^3(1-x)^2 \quad \text{or} \quad 10x^3(1-x)^2.$$

Testing for independence

In the examples we have considered so far, it has been obvious whether or not two events A and B are independent. In particular, A and B clearly *are* independent when they are defined with reference to physically distinct experiments such as drawing a card from a pack and tossing a coin. The independence is not always so obvious, however. Suppose, for example, that a die is thrown twice, and A and B are the events 'the total is 10 or more' and 'the score on the second die is 4'. Here it is not immediately clear whether the occurrence of one of the events affects the probability of the other, and we therefore need a method of testing for independence.

The method of testing is simply to see whether or not $p(A \text{ and } B) = p(A) \times p(B)$. In the above example it is easily shown that $p(A) = \frac{1}{6}$, $p(B) = \frac{1}{6}$ and $p(A \text{ and } B) = \frac{1}{36}$; so the events *are* independent. The following two examples are slightly more difficult and involve the use of the formula for $p(A \cup B)$.

Example 7 Determine whether the events A and B are independent, given that $p(A) = \frac{2}{3}$, $p(B) = \frac{1}{4}$ and $p(A \cup B) = \frac{17}{24}$.

We need to find $p(A \cap B)$. Using the theorem

$$p(A \cup B) = p(A) + p(B) - p(A \cap B),$$

we have

$$\frac{17}{24} = \frac{2}{3} + \frac{1}{4} - p(A \cap B)$$

$$\therefore \; p(A \cap B) = \frac{16}{24} + \frac{6}{24} - \frac{17}{24}$$

$$= \frac{5}{24}.$$

Since $p(A) \times p(B) = \frac{1}{6}$, which is not equal to $\frac{5}{24}$ and thus to $p(A \cap B)$, the events A and B are **not independent**.

Example 8 In a certain country 80% of the adults have televisions, 24% have televisions and are short-sighted, and 86% either have televisions or are short-sighted or have both of these characteristics. Find the probability that an adult chosen at random is short-sighted, and determine whether possession of a television and short-sightedness are independent.

Denoting possession of a television by T and short-sightedness by S, we have $p(T) = \frac{40}{50}$, $p(T \cap S) = \frac{12}{50}$ and $p(T \cup S) = \frac{43}{50}$. Hence since

$$p(T \cup S) = p(T) + p(S) - p(T \cap S),$$

we have

$$\frac{43}{50} = \frac{40}{50} + p(S) - \frac{12}{50}$$

$$\therefore \ p(S) = \frac{15}{50} = \frac{3}{10}.$$

Now since $p(T) \times p(S) = \frac{4}{5} \times \frac{3}{10} = \frac{6}{25} = 24\% = p(T \cap S)$, it follows that possession of a television and short-sightedness are **independent**.

Exercise 15b

1   A coin and a die are thrown. Find the probability of (a) a head, a 6 or both, (b) a tail, an odd number or both.

2   A man plays two games of darts, his probability of a win being $\frac{2}{3}$ in each game. Find the probability that he wins in either or both of the games.

3   A card is drawn from an ordinary pack and a die is thrown. Find the probability of (a) a diamond or a 2 or both, (b) an ace, an odd number or both.

4 A and B are independent events. Find $p(A \cup B)$, given that (a) $p(A) = \frac{3}{4}$ and $p(B) = \frac{2}{5}$, (b) $p(B) = \frac{1}{3}$ and $p(A \cap B) = \frac{2}{15}$.

5   Given that $p(A) = \frac{1}{5}$, $p(B) = \frac{3}{10}$ and $p(A \cup B) = \frac{11}{25}$, find $p(A \cap B)$ and determine whether A and B are independent.

6 A and B are mutually exclusive events. (a) Find $p(A \cup B)$, given that $p(A) = \frac{2}{3}$ and $p(B) = \frac{1}{6}$. (b) Find $p(A)$, given that $p(B') = \frac{3}{8}$ and $p(A \cup B) = \frac{13}{16}$.

7   A man and his wife play in a tournament, their chances of winning being $\frac{5}{12}$ and $\frac{1}{8}$, respectively. Find the probability that either of them wins. (The tournament has only one winner.)

8   Sarah and Judy run in a race. If the probability that either Sarah wins or Judy wins is $\frac{13}{15}$, and the probability that Sarah loses is $\frac{3}{5}$, what is the probability that Judy loses?

9   A die is thrown once. Determine whether the events 'getting an odd number' and 'getting a 1 or a 4' are independent.

10   A number is chosen at random from the set {natural numbers from 5 to 20 inclusive}. In each of the following cases determine whether the given two events are independent.
(a)   The number is prime. The number is over 17.
(b)   The number is even. The number is under 15.
(c)   The number is 9, 10 or 11. The number is 10, 12 or 14.

11   A card is drawn at random from an ordinary pack and a die is thrown. Find the probability of either a heart followed by an odd number or a black card followed by a 6.

12   Two balls are drawn at random from a bag containing 4 red, 3 green and 2 blue balls. Find the probability that either a red ball is followed by a blue or a green ball is followed by a red.

13   A chess player's probability of a win is $\frac{2}{5}$ if he has White, and $\frac{1}{2}$ if he has Black.

What is his overall probability of a win if the colours are determined by tossing a coin?

14 What is the probability that (a) a family of 3 children contains exactly 2 boys, (b) a family of 4 children contains exactly 2 boys, (c) a family of 5 children contains exactly 3 boys?

15 My probability of getting to work on time is $\frac{9}{10}$ if I drive, and $\frac{4}{5}$ if I go by other means. Given that I drive on 2 days out of 3, calculate my overall probability of getting to work on time.

16 A marksman hits the target, on average, 3 times in every 4 shots. What is the probability that (a) in 2 shots he has exactly 1 hit, (b) in 4 shots he has exactly 3 hits?

17 In each of the following cases determine whether the events A and B are independent.

(a) $p(A) = \frac{5}{12}$, $p(B) = \frac{9}{20}$, $p(A \cap B) = \frac{5}{16}$.

(b) $p(A) = \frac{5}{6}$, $p(B) = \frac{7}{9}$, $p(A \cup B) = \frac{26}{27}$.

(c) $p(A) = 55\%$, $p(A \cap B) = 22\%$, $p(A \cup B) = 73\%$.

(d) $p(A') = \frac{7}{12}$, $p(B') = \frac{1}{5}$, $p(A \cup B) = \frac{9}{10}$.

18 In a certain school 60% of the pupils like music, 20% like chess and 65% like music, chess or both. Find the probability that a pupil chosen at random likes both music and chess, and determine whether liking music and liking chess are independent.

19 If it rains, on average, on one day in every 3, what is the probability that in a period of 4 days it rains on fewer than 2 days?

20 Two events each have a probability of p. Find p, given that the probability that neither event occurs is 0.15 and the probability that both occur is 0.35.

21 Two dice are thrown. In each of the following cases determine whether the given two events are independent.

(a) The total is 7. The score on the second die is 4.

(b) The total is less than 4. The scores on the dice differ.

(c) The first die scores more than the second. The score on the second die is 3 or 4.

22 If 10% of eggs are cracked, what is the probability, to 2 d.p., that in a box of 6 eggs I get more than 1 cracked egg?

23 A survey of 1500 people shows that 1200 are healthy, 500 exercise regularly and 1300 have either or both of these characteristics. Determine whether being healthy and taking regular exercise are independent.

24 Two cards are drawn at random from a reduced pack consisting of the aces and kings of each suit. Calculate the probability that (a) the first card is an ace and the second is a heart, (b) the first card is a red ace and the second is a black king, (c) the first card is red and the second is a red king. In each case say whether the two events are independent.

Tree diagrams

Tree diagrams are useful for dealing with experiments and situations in which many different combinations of outcomes are possible. Suppose, for example, that two balls are drawn at random, without replacement, from a bag containing 4 white balls, 3 black balls and 1 green ball. The following tree diagram (Fig. 15.4) shows all the possible outcomes and their probabilities.

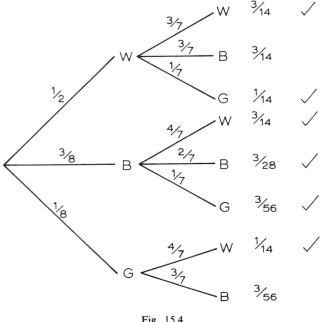

Fig. 15.4

The following points should be noted.

1. The branches leading away from any 'node' represent mutually exclusive events and their probabilities add up to 1.

2. Each fraction on the extreme right is the probability of the complete set of branches leading to it, and is obtained by multiplying the probabilities on the individual branches. For example, the first fraction, $\frac{3}{14}$, is the probability that the two balls drawn are both white, and it is obtained by multiplying $\frac{1}{2}$ by $\frac{3}{7}$.

3. The fractions on the extreme right are probabilities of mutually exclusive events, and hence they should add up to 1. This provides a simple check on the correctness of the tree diagram. *This check should always be carried out before a tree diagram is used to solve problems.*

A tree diagram is particularly helpful when the event whose probability is required can occur in several different ways, so that several sets of branches of the tree diagram contribute to the event. With reference to the above experiment, for example, consider the event 'a black ball is drawn first or not at all.' There are 6 sets of branches which contribute to this event (ticked, in Fig. 15.4), and since these represent 6 mutually exclusive sets of outcomes, we obtain the required probability by adding their individual probabilities:

$$\frac{3}{14}+\frac{1}{14}+\frac{3}{14}+\frac{3}{28}+\frac{3}{56}+\frac{1}{14}=\frac{41}{56}.$$

Conditional probability

It was shown above that when two events A and B are not independent, we have

$$p(A \text{ and } B) = p(A) \times p(B|A)$$
$$\text{or} \quad p(A \text{ and } B) = p(B) \times p(A|B).$$

In some problems a conditional probability itself has to be calculated, and hence a formula is required in which the conditional probability is the subject. From the second of the above formulae, we have

$$p(A|B) = \frac{p(A \text{ and } B)}{p(B)}.$$

The first of the following examples shows how this formula can be used when the given information is displayed in the form of a tree diagram.

Example 9 A box contains 1 black ball and 1 white ball. After a draw, the ball drawn is replaced in the box together with a ball of the other colour. Draw a tree diagram for 3 random draws and use it to find the probability that (a) the second and third balls drawn have the same colour (event A), (b) exactly one of the balls drawn is white (event B). Find also (c) $p(A|B)$, (d) $p(B|A)$.

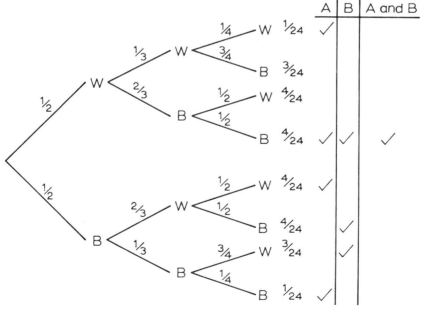

Fig. 15.5

(Note that the fractions on the right are all given with the same denominator. This assists the calculations. Note also that these fractions add up to 1.)

(a) The ticks in the column headed A indicate the sets of branches which contribute to the event A. We have

$$p(A) = \frac{1}{24} + \frac{4}{24} + \frac{4}{24} + \frac{1}{24} = \frac{10}{24} = \frac{5}{12}.$$

(b) Similarly, using the ticks in column B, we obtain

$$p(B) = \frac{4}{24} + \frac{4}{24} + \frac{3}{24} = \frac{11}{24}.$$

(c) Using the formula

$$p(A|B) = \frac{p(A \text{ and } B)}{p(B)},$$

we have

$$p(A|B) = \frac{4/24}{11/24} = \frac{4}{11}.$$

(d) Using the formula

$$p(B|A) = \frac{p(B \text{ and } A)}{p(A)},$$

we have

$$p(B|A) = \frac{4/24}{10/24} = \frac{2}{5}.$$

Example 10   My probability of promotion is $\frac{1}{3}$, and if I am promoted the probability that I change my car is $\frac{3}{4}$. If I am not promoted the probability that I change my car is $\frac{1}{2}$. Find the probability that I am promoted, given that I do not change my car.

Denoting promotion by P and a change of car by C, we require $p(P|\text{not-}C)$, that is

$$\frac{p(P \text{ and not-}C)}{p(\text{not-}C)}.$$

In this case the numerator can be obtained immediately from the given data. We have

$$p(P \text{ and not-}C) = \frac{1}{3} \times \frac{1}{4} = \frac{1}{12}.$$

The denominator is the more difficult to calculate since the event not-C is the union of the two mutually exclusive events

$$P \text{ and not-}C, \quad \text{not-}P \text{ and not-}C.$$

It follows that $p(\text{not-}C)$ is the sum of the probabilities of these two events. We have

$$p(P \text{ and not-}C) = \frac{1}{3} \times \frac{1}{4} = \frac{1}{12},$$

$$p(\text{not-}P \text{ and not-}C) = \frac{2}{3} \times \frac{1}{2} = \frac{1}{3}.$$

It now follows that

$$p\,(\text{not-}C) = \frac{1}{12} + \frac{1}{3} = \frac{5}{12}$$

and finally we have

$$p\,(P\,|\,\text{not-}C) = \frac{1/12}{5/12} = \frac{1}{5}.$$

Exercise 15c

1   Two balls are drawn at random, without replacement, from a bag containing 5 red balls, 3 white balls and 2 blue balls. Draw a tree diagram and find the probability that (a) exactly 1 white ball is drawn, (b) the two balls drawn have the same colour, (c) 4 red balls, 3 white balls and 1 blue ball are left in the bag.

2   All days are classed as 'fine' or 'wet'. If one day is fine the probability that the next day is fine is 0.7, while if one day is wet the probability that the next is wet is 0.4. Given that Sunday is wet, draw a tree diagram for Monday, Tuesday and Wednesday, and find the probability that (a) Wednesday is fine, (b) Monday and Wednesday have the same weather.

3   A woman with six 10p coins and two 5p coins in her pocket draws out three coins at random without replacement. Draw a tree diagram and find the probability that (a) she draws 25p, (b) the third coin she draws differs from the first.

4   Two boxes contain, respectively, 3 aces, 2 kings and 1 queen, and 2 aces, 1 king and 3 queens. A man chooses a box and draws one random card from it. Given that he is twice as likely to choose the first box as the second, draw a tree diagram and find the probability that the card drawn is (a) an ace, (b) either a king from the first box or a queen from the second.

5   Ann and Mary play four games of draughts. After Ann wins her probability of winning the next game is $\frac{4}{5}$, and after Mary wins her probability of winning the next game is $\frac{3}{5}$. Given that Ann wins the first game, draw a tree diagram for the next three games and find the probability that (a) Mary wins the fourth game, (b) the score after the four games is 2-all, (c) Ann wins at least three games. (No games are drawn.)

6   A bag initially contains 3 white balls and 3 black balls. After a ball is drawn it is replaced by one of the other colour. Draw a tree diagram for three random draws and find the probability that (a) the last two balls drawn have the same colour, (b) the balls drawn alternate in colour, (c) the bag finally contains 2 white and 4 black balls.

7   Two dice are thrown. Find the probability that (a) the total is more than 7, given that the second die scores a 3 or a 4, (b) the second die scores a 3 or a 4, given that the total is more than 7.

8   A coin is thrown three times. Find the probability that (a) a head is obtained on the third throw, given that exactly 2 heads are obtained, (b) exactly 2 tails are obtained, given that a tail is obtained on the second throw.

9   With the data of question 1, find the probability that (a) the first ball drawn is blue, given that the second is not blue, (b) both balls drawn are red, given that at least one of them is red.

10   In a country containing equal numbers of men and women, $\frac{3}{4}$ of the men and

$\frac{1}{3}$ of the women drive cars. Find the probability that a person chosen at random is a man, given that he or she drives a car.

11   With the data of question 4, find the probability that (a) the card is drawn from the first box, given that it is an ace, (b) it is drawn from the second box, given that it is not a king.

12   With the data of question 3, find the probability that (a) the woman draws a total of 20p, given that the second coin she draws is a 10p, (b) the first coin differs from the second, given that the second differs from the third.

13   If it rains the probability that I go to work by bus is $\frac{5}{6}$, while if it does not rain the probability that I go by bus is $\frac{2}{7}$. It rains, on average, on 3 days out of every 10. Find the probability that it rains on a certain day, given that I go to work by bus on that day.

14   When John plays his friend at chess, his probability of a win, a loss and a draw are respectively 0.4, 0.5 and 0.1. Given that John is ahead after they have played three games, find the probability that none of the games were drawn.

16

General Kinematics of a Particle in a Plane

There are certain types of motion – e.g. motion with constant acceleration – which occur so frequently that it is convenient to derive standard equations for them relating displacement, velocity, acceleration and time. In this chapter we consider motion in a more general way, and show how different kinds of motion can be analysed merely by the use of the definitions of acceleration (a), velocity (v) and displacement (s).

Motion in a straight line

The quantities s, v and a are of course vectors, possessing direction as well as magnitude. As was pointed out in Chapter 1, however, direction can be indicated by the use of plus and minus signs when the motion is in a straight line. It is therefore unnecessary in these circumstances to print the symbols s, v and a in bold type or to underline them.

The general analysis of motion requires the use of calculus, and in particular the following relationships:

$$v = \frac{ds}{dt}, \quad a = \frac{dv}{dt}, \quad a = v\frac{dv}{ds}.$$

The first two of these are definitions, and the last can be proved immediately by writing

$$\frac{dv}{dt} = \frac{dv}{ds}\frac{ds}{dt} = \frac{dv}{ds}v.$$

The usual starting point for the analysis of a motion is an expression for *acceleration* in terms of either t, s or v. This is because we are normally told or can find the forces acting on the moving body and can then deduce the acceleration by using the equation $F = ma$. Expressions for v and s can be obtained by integrating the expression for a. It is always necessary to decide, before integrating, whether to denote the acceleration by dv/dt or by $v\, dv/ds$. The following rules may be helpful:

a given in terms of t: use $\dfrac{dv}{dt}$;

a given in terms of *s*: use $v\dfrac{dv}{ds}$;

a given in terms of *v*: use $\dfrac{dv}{dt}$ if *v* is required in terms of *t*;

use $v\dfrac{dv}{ds}$ if *v* is required in terms of *s*.

Worked examples

Note: In kinematics problems it is essential *always* to put in the arbitrary constant after integrating. The value of the constant should be calculated from the information given, and the calculation should be shown even when the constant is zero.

In the examples which follow the units are SI (metres, seconds, newtons, etc.) throughout, and all the motion is in a straight line.

Example 1   A particle starts from rest and moves with an acceleration of cos 2*t*. Find its velocity and displacement from the starting point after a time of $\pi/6$ s.

We have

$$\frac{dv}{dt} = \cos 2t$$

$$\therefore \; v = \tfrac{1}{2}\sin 2t + c.$$

Since $v = 0$ when $t = 0$, it follows that $c = 0$. Hence

$$v = \tfrac{1}{2}\sin 2t.$$

When $t = \pi/6$,

$$v = \tfrac{1}{2}\sin (\pi/3)$$

$$= \sqrt{3/4}\,\mathbf{m\,s^{-1}}.$$

To obtain an equation for *s* we write *v* as d*s*/d*t* and integrate again:

$$\frac{ds}{dt} = \tfrac{1}{2}\sin 2t$$

$$\therefore \; s = -\tfrac{1}{4}\cos 2t + k.$$

Since *s* is measured from the starting point, $s = 0$ when $t = 0$. Hence

$$0 = -\tfrac{1}{4} + k$$

$$\therefore \; k = \tfrac{1}{4}$$

$$\therefore \; s = \tfrac{1}{4}(1 - \cos 2t).$$

When $t = \pi/6$,

$$s = \tfrac{1}{4}(1 - \tfrac{1}{2})$$

$$= \tfrac{1}{8}\,\mathbf{m}.$$

Example 2   A particle starts at the point (4,0) with a velocity of $3\,\mathrm{m\,s^{-1}}$, and moves along the *x*-axis with a constant acceleration towards the origin of $2\,\mathrm{m\,s^{-2}}$.

Find (a) expressions for v and x in terms of t, (b) the maximum positive distance from the origin, (c) the time the particle takes to reach the origin.

(In this type of question it is convenient to use x for the displacement from the origin, rather than s. We shall assume that the x-axis is marked out in metres.)
(a) We have

$$\frac{dv}{dt} = -2$$

$$\therefore v = -2t + c.$$

Since we are given that $v = 3$ when $t = 0$, $c = 3$ and hence

$$v = -2t + 3.$$

To obtain s we write v as dx/dt and integrate again:

$$\frac{dx}{dt} = -2t + 3$$

$$\therefore x = -t^2 + 3t + k.$$

Since the particle starts at $(4, 0)$, $x = 4$ when $t = 0$ and thus $k = 4$. Hence

$$x = -t^2 + 3t + 4.$$

(b) The maximum value of x occurs when $dx/dt = 0$, i.e. when $v = 0$. Hence

$$-2t + 3 = 0$$

$$\therefore t = \tfrac{3}{2}$$

$$\therefore x_{max} = -\tfrac{9}{4} + \tfrac{9}{2} + 4.$$

$$= 6\tfrac{1}{4}\,\text{m}.$$

(c) When the particle is at the origin, $x = 0$ and we have

$$-t^2 + 3t + 4 = 0$$

$$\therefore t^2 - 3t - 4 = 0$$

$$\therefore (t - 4)t + 1) = 0$$

$$\therefore t = 4 \quad \text{or} \quad -1.$$

The solution -1 corresponds to a time before the motion started. This solution must therefore be discarded and it follows that the required time is **4 s**.

Example 3 A body of mass m starts with a velocity of V_0 and moves against a retarding force of kv. Obtain expressions for v in terms of s and in terms of t. (Both s and t are measured from the starting point.)

Using $F = ma$, and remembering that we have a *retarding* force, we obtain $a = -kv/m$. If we require v in terms of s we express the acceleration as $v\,dv/ds$.

$$v\frac{dv}{ds} = -\frac{kv}{m}$$

$$\therefore \frac{dv}{ds} = -\frac{k}{m}.$$

Integrating now with respect to s, we have

$$v = -\frac{ks}{m} + c.$$

Since $v = V_0$ when $s = 0$, $c = V_0$ and hence

$$v = V_0 - \frac{ks}{m}.$$

To obtain v in terms of t we express the acceleration as dv/dt.

$$\frac{dv}{dt} = -\frac{kv}{m}.$$

Separating the variables, we have

$$\frac{dv}{v} = -\frac{k\,dt}{m}.$$

Integrating both sides,

$$\ln v = -\frac{kt}{m} + c_1 \qquad \text{(where } \ln v = \log_e v\text{)}.$$

Since $v = V_0$ when $t = 0$, $c_1 = \ln V_0$. Hence

$$\ln v = -\frac{kt}{m} + \ln V_0$$

$$\therefore \ln \frac{v}{V_0} = -\frac{kt}{m}$$

$$\therefore \frac{v}{V_0} = e^{-kt/m}$$

$$\therefore v = V_0 e^{-kt/m}.$$

Example 4   A body is projected vertically upwards with a speed of $10\,\mathrm{m\,s^{-1}}$, and in addition to its own weight experiences a retarding force which is proportional to the square of its speed. The initial value of this force is half the body's weight. Find the maximum height of the body above its starting point.

When the body is moving upwards at $v\,\mathrm{m\,s^{-1}}$ there are two downward forces on it, as shown in Fig. 16.1. To find the constant of proportionality k, we use the fact that $kv^2 = mg/2 = 5m$ when $v = 10$:

$$100k = 5m$$

$$\therefore k = \frac{m}{20}.$$

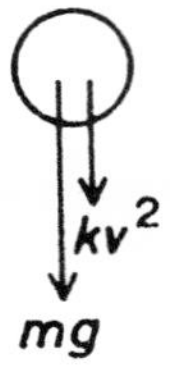

$$kv^2$$

$$mg$$

Fig. 16.1

Now using $F = ma$, we have

$$\text{upward acceleration} = -\left(\frac{mv^2}{20} + 10m\right) \div m$$

that is,

$$v\frac{dv}{ds} = -\frac{v^2 + 200}{20}.$$

Separating the variables,

$$\frac{v\,dv}{v^2 + 200} = -\frac{ds}{20}.$$

Integrating both sides,

$$\tfrac{1}{2}\ln(v^2 + 200) = -\frac{s}{20} + c.$$

Since $v = 10$ when $s = 0$, $c = \tfrac{1}{2}\ln 300$; hence

$$\tfrac{1}{2}\ln(v^2 + 200) - \tfrac{1}{2}\ln 300 = -\frac{s}{20}$$

$$\therefore\ \tfrac{1}{2}\ln\frac{v^2 + 200}{300} = -\frac{s}{20}.$$

The maximum height occurs when $v = 0$, i.e. when

$$\tfrac{1}{2}\ln\tfrac{2}{3} = -\frac{s}{20}.$$

Hence

$$s_{max} = -10\ln\tfrac{2}{3}$$
$$= \mathbf{10\ln\tfrac{3}{2}\ metres.}$$

Exercise 16a

All the motion is in a straight line. Where they are not explicitly stated, the units are SI. Displacement s is always measured from the point at which $t = 0$.

1 A particle starts with a velocity of $2\,\mathrm{m\,s^{-1}}$, and moves with an acceleration of $4t$. Find its velocity and displacement after $2\,\mathrm{s}$.

2 A particle starts at rest and moves with an acceleration of $\sin 2t$. Find its velocity after a time of $\pi/3\,\mathrm{s}$.

3 A particle starts at the point $(3, 0)$ with a velocity of $4\,\mathrm{m\,s^{-1}}$, and moves along the x-axis with an acceleration towards the origin of $2t$. Find (a) its greatest positive distance from the origin, (b) the time it takes to return to its starting point.

4 The velocity of a particle is given by e^{-2s}. Express s in terms of t.

5 A particle starts with a velocity of V_0 and moves with an acceleration of $2v^2$. Prove that $v = V_0 e^{2s}$.

6 A particle moves with a velocity given by $6t - 2$. Find its velocity when its displacement from the starting point is $1\,\mathrm{m}$.

7 A particle moves according to the equation $s = 1 + t^2 - t^3$. Find the maximum positive values of s and v.

8 A particle starts at rest and moves with an acceleration of $6 - 2t$. Find its maximum velocity.

9   A particle starts with a velocity of $1\,\mathrm{m\,s^{-1}}$, and moves in such a way that its velocity always equals its acceleration. Express (a) v in terms of s, (b) v in terms of t, (c) s in terms of t, (d) t in terms of s.

10   A particle moves with a velocity given by $v = t + t^2 - t^3$. Find its maximum velocity and its maximum acceleration.

11   A particle starts at rest and moves with an acceleration of $9/v^2$. Find its velocity and displacement after $8\,\mathrm{s}$.

12   A particle moves according to the equation $s = \sin 2t - \cos 2t$. Prove that $a = -4s$ and $v^2 = 4(2 - s^2)$.

13   A particle starts with a velocity of $3\,\mathrm{m\,s^{-1}}$ and moves with an acceleration of $2v$. Express (a) v in terms of s, (b) s in terms of t.

14   A particle starts with a velocity of $\pi/4$, and moves with an acceleration of $\cos^2 v$. Prove that $v = \arctan(t + 1)$.

15   A particle moves according to the equation $s = \sin 4t$. Find the maximum positive values of s, v and a.

16   A particle starts at $(2, 0)$ with a velocity of $3\,\mathrm{m\,s^{-1}}$, and moves along the x-axis with an acceleration which varies directly as the velocity. The initial acceleration is $6\,\mathrm{m\,s^{-2}}$. Express x in terms of t.

17   A $4\,\mathrm{kg}$ body moves under the influence of a retarding force which is proportional to its velocity. Its initial velocity is $8\,\mathrm{m\,s^{-1}}$, and the initial value of the force is $64\,\mathrm{N}$. Obtain expressions for v and s in terms of t, and find the distance travelled when the initial velocity is halved.

18   A particle moves according to the equation $v = 2 - s$. Prove that $s = 2 - 2e^{-t}$ and show that the body is halfway to its limiting position after a time of $\ln 2$.

19   A body of mass m is projected vertically upwards with a speed of $20\,\mathrm{m\,s^{-1}}$, and in addition to its own weight experiences a retarding force of $mv/2$. Prove that $v = 20(2e^{-t/2} - 1)$ and deduce that the body reaches its maximum height after a time of $\ln 4$.

20   A $2\,\mathrm{kg}$ body starts at rest and moves under the influence of a constant driving force of $20\,\mathrm{N}$ and a retarding force which is proportional to the square of the time in seconds from the starting point. If the acceleration is $4\,\mathrm{m\,s^{-2}}$ after $1\,\mathrm{s}$, how far does the body travel before coming to rest again?

21   A $4\,\mathrm{kg}$ body is projected vertically upwards with a speed of $5\,\mathrm{m\,s^{-1}}$, and in addition to its own weight experiences a retarding force of $8v^2$. Prove that when its height in metres is h, $v^2 = 30e^{-4h} - 5$, and deduce that its maximum height is $\frac{1}{4}\ln 6$.

22   A body is released from rest in a viscous medium which provides a resistance which is proportional to the square of the velocity. Show that if the body approaches a constant terminal velocity of $10\,\mathrm{m\,s^{-1}}$, then this resistance can be expressed as $mv^2/10$, where m is the mass of the body in kilograms. Hence show that when the body has fallen through h metres, $v^2 = 100(1 - e^{-h/5})$, and deduce that half of the terminal velocity is reached when $h = 5\ln 1\frac{1}{3}$.

Motion in two dimensions

When a particle moves in a plane curve it is convenient to introduce x and y axes and analyse its motion with the aid of the unit vectors $\mathbf{i}$ and $\mathbf{j}$. We begin by showing that velocity and acceleration, expressed in terms of $\mathbf{i}$ and $\mathbf{j}$, can be obtained by differentiating the expression for displacement, and thus that the definitions given in the last section for motion in a straight line continue to apply when the motion is in a curve.

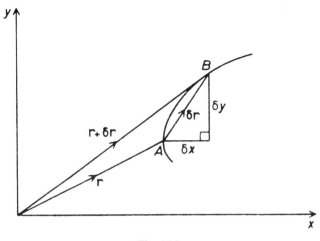

Fig. 16.2

Consider a particle moving along a path in the x–y plane, as shown in Fig. 16.2, and suppose that in a time of δt it moves a small distance from $A(x, y)$ to $B(x + \delta x, y + \delta y)$. The vector $\overrightarrow{AB}$ then represents the increase $\delta\mathbf{r}$ in the displacement $\mathbf{r}$ from the origin. By the triangle law, we have

$$\delta\mathbf{r} = \delta x\mathbf{i} + \delta y\mathbf{j}$$

$$\therefore \quad \frac{\delta\mathbf{r}}{\delta t} = \frac{\delta x}{\delta t}\mathbf{i} + \frac{\delta y}{\delta t}\mathbf{j}.$$

Now let the point B approach A, so that $\delta\mathbf{r}$, δx, etc. all approach zero. The magnitude of the vector $\delta\mathbf{r}$ then approaches the length of arc AB, and its direction approaches that of the tangent at A. It follows that the limit $d\mathbf{r}/dt$ must represent the velocity of the particle at A, and we have

$$\mathbf{v} = \frac{d\mathbf{r}}{dt} = \frac{dx}{dt}\mathbf{i} + \frac{dy}{dt}\mathbf{j}.$$

In a similar way it can be shown that the acceleration vector, $\mathbf{a}$, is given by

$$\mathbf{a} = \frac{d\mathbf{v}}{dt} = \frac{d^2 x}{dt^2}\mathbf{i} + \frac{d^2 y}{dt^2}\mathbf{j}.$$

Example If a particle moves in the x–y plane in such a way that $\mathbf{r} = t^2\mathbf{i} - t^3\mathbf{j}$, we have $\mathbf{v} = 2t\mathbf{i} - 3t^2\mathbf{j}$ and $\mathbf{a} = 2\mathbf{i} - 6t\mathbf{j}$.

Equation of path

As in the case of projectiles (see Chapter 13) the equation of the path of a particle moving in a plane is obtained by eliminating t from the equations for x and y. Thus if, for example,

$$\mathbf{r} = 2t\mathbf{i} - t^2\mathbf{j},$$

we have

$$x = 2t, \qquad y = -t^2,$$

from which the equation of the path is

$$4y + x^2 = 0.$$

Worked examples

Example 1　A particle moves in the x–y plane with an acceleration which is given by $\mathbf{a} = (6t - 2)\mathbf{i} + 2\mathbf{j}$. Its initial displacement from the origin is given by $\mathbf{r} = 2\mathbf{i} + 3\mathbf{j}$, and its initial velocity is $-2\mathbf{j}$. Obtain expressions for $\mathbf{v}$ and $\mathbf{r}$ in terms of t and find the magnitude and direction of its velocity after 2 s.

We have

$$\mathbf{a} = (6t - 2)\mathbf{i} + 2\mathbf{j}.$$

Hence, integrating,

$$\mathbf{v} = (3t^2 - 2t)\mathbf{i} + 2t\mathbf{j} + c.$$

Since $\mathbf{v} = -2\mathbf{j}$ when $t = 0$, $c = -2\mathbf{j}$ and we thus have

$$\mathbf{v} = (3t^2 - 2t)\mathbf{i} + (2t - 2)\mathbf{j}.$$

Integrating again,

$$\mathbf{r} = (t^3 - t^2)\mathbf{i} + (t^2 - 2t)\mathbf{j} + k.$$

When $t = 0$, $\mathbf{r} = 2\mathbf{i} + 3\mathbf{j}$; hence $k = 2\mathbf{i} + 3\mathbf{j}$ and we have

$$\mathbf{r} = (t^3 - t^2 + 2)\mathbf{i} + (t^2 - 2t + 3)\mathbf{j}.$$

When $t = 2$,

$$\mathbf{v} = 8\mathbf{i} + 2\mathbf{j}.$$

Hence

$$|\mathbf{v}| = \sqrt{8^2 + 2^2}$$
$$= \sqrt{68}$$
$$= 2\sqrt{17} \, \mathbf{m \, s^{-1}}.$$

(The magnitude of the velocity, $|\mathbf{v}|$, is of course the same as the speed.)
　　The velocity is directed at θ to the positive x-axis, where

$$\tan \theta = \tfrac{2}{8} = \tfrac{1}{4}.$$

Hence $\theta = \mathbf{arc \, tan} \tfrac{1}{4}$.

Example 2　A particle P starts at $(a, 0)$ and moves in an anticlockwise circle round the origin at a constant angular speed of 2 rad s^{-1}. Obtain an expression for $\mathbf{r}$ in terms of t and derive an expression for the acceleration, $\mathbf{a}$. Hence show that $\mathbf{a}$ is always directed towards the centre and of magnitude $4a$. If another particle Q starts at the same point and moves in a clockwise circle at 4 rad s^{-1}, what is the velocity of Q relative to P after $\pi/6$ s?

P moves in a circle of radius *a*. Since its angular speed is 2 rad s^{-1}, it turns through an angle of 2*t* radians in *t* seconds (Fig. 16.3).

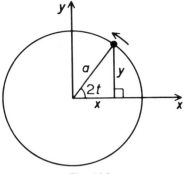

Fig. 16.3

From the diagram,
$$x = a \cos 2t \quad \text{and} \quad y = a \sin 2t,$$
and we thus have
$$\mathbf{r} = a \cos 2t\,\mathbf{i} + a \sin 2t\,\mathbf{j}.$$
Differentiating twice,
$$\mathbf{v} = -2a \sin 2t\,\mathbf{i} + 2a \cos 2t\,\mathbf{j},$$
$$\mathbf{a} = -4a \cos 2t\,\mathbf{i} - 4a \sin 2t\,\mathbf{j}.$$

We can now find the magnitude and direction of **a** by drawing a parallelogram of vectors (Fig. 16.4).

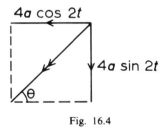

Fig. 16.4

By Pythagoras
$$|\mathbf{a}| = \sqrt{16a^2 \cos^2 2t + 16a^2 \sin^2 2t}$$
$$= 4a \quad (\text{since } \sin^2 x + \cos^2 x = 1).$$
Also
$$\tan \theta = \frac{4a \sin 2t}{4a \cos 2t} = \tan 2t.$$

Hence $\theta = 2t$ and the direction of **a** is towards the centre.

The second particle *Q* turns through $-4t$ radians in *t* seconds (regarding a clockwise angle as negative, in the usual way) and, reasoning as above, we thus have for this particle
$$\mathbf{r} = a \cos 4t\,\mathbf{i} - a \sin 4t\,\mathbf{j}$$
$$\therefore \ \mathbf{v} = -4a \sin 4t\,\mathbf{i} - 4a \cos 4t\,\mathbf{j}.$$

Substituting $t = \pi/6$ into both of the expressions for **v**, we now obtain

$$\text{velocity of } P = -2a\frac{\sqrt{3}}{2}\mathbf{i} + 2a \times \tfrac{1}{2}\mathbf{j}$$

$$= -a\sqrt{3}\mathbf{i} + a\mathbf{j},$$

$$\text{velocity of } Q = -4a\frac{\sqrt{3}}{2}\mathbf{i} + 4a \times \tfrac{1}{2}\mathbf{j}$$

$$= -2a\sqrt{3}\mathbf{i} + 2a\mathbf{j}.$$

Hence

velocity of Q relative to P = velocity of Q − velocity of P

$$= -a\sqrt{3}\mathbf{i} + a\mathbf{j}.$$

Exercise 16b

Displacement is measured from the origin and denoted by **r**. All motion is in the x–y plane and all units are SI.

1   A particle moves such that its displacement at time t is given by $\mathbf{r} = t^3\mathbf{i} - 2t\mathbf{j}$. Obtain vector expressions for (a) the velocity and acceleration at time t, (b) the velocity and acceleration when $t = 2$.

2   A particle starts at $(2,3)$ and moves according to the equation $\mathbf{v} = 2t\mathbf{i} + t^2\mathbf{j}$. Find (a) an expression for **r** in terms of t, (b) the position when $t = 3$, (c) the position when the particle is moving at $45°$ to the positive x-axis.

3   A particle starts at the origin and moves according to the equation $\mathbf{v} = 2\mathbf{i} - t\mathbf{j}$. Find the equation of the path.

4   A particle starts at a point on the x-axis with a velocity of $-3\mathbf{j}$, and moves with an acceleration of $(1 - 2t)\mathbf{i} + 2\mathbf{j}$. Find the speed of the particle and the angle between its direction and the positive x-axis at the moment when it reaches the x-axis for the second time.

5   A particle starts at the origin with a velocity of $2\mathbf{i}$ and moves with an acceleration of $3\mathbf{j}$. Obtain the equation of the path. Show that the particle passes through the point $(4,6)$, and find its speed at this point.

6   A particle starts at the origin with a velocity of $4\mathbf{i}$ and moves with an acceleration of $-\mathbf{i} + 2t\mathbf{j}$. Find its position and speed when the angle between its direction and the positive x-axis is arc tan 2.

7   Two particles move such that their displacements are given by

$$\mathbf{r} = t^2\mathbf{i} + 2t\mathbf{j} \quad \text{and} \quad \mathbf{r} = (t - 1)\mathbf{i} + t\mathbf{j}.$$

Find (a) the point through which both particles pass and time interval between the moments at which they reach it, (b) the velocity of the first particle relative to the second after 2 s.

8   A particle starts at the point $(-4,1)$ with a velocity of $-2\mathbf{j}$, and moves with a constant acceleration of $6\mathbf{i}$. Find (a) its position and speed after 1 s, (b) the times required for it to reach the x and y axes, respectively.

9   One particle starts at the origin with a velocity of $6\mathbf{j}$ and moves with an acceleration of $-2\mathbf{j}$, while another moves according to the equation $\mathbf{r} = (4 - t)\mathbf{i} + kt\mathbf{j}$. What is the value of k if they collide?

10   A particle moves according to the equation $\mathbf{r} = \sin 2t\,\mathbf{i} + \cos 2t\,\mathbf{j}$. Prove (a) that it moves in a circle, centre the origin, with a constant speed of 2, (b) that the acceleration is always directed towards the centre and of magnitude 4.

11   A particle starts at the point $(0,k)$ and moves according to the equation $\mathbf{v} = 4t\mathbf{i} - \mathbf{j}$. Find k if it passes through $(8,3)$.

12   A particle starts at rest at the point $(0,3)$ and moves with an acceleration of $4\mathbf{i} + 8\mathbf{j}$. Show that it moves along the line $y = 2x + 3$ and find (a) the distance it travels in the first second, (b) the distance it has travelled when its speed is $12\sqrt{5}$.

13   Two particles P and Q move according to the equations $\mathbf{r} = (t-1)\mathbf{i}$ and $\mathbf{r} = t^2\mathbf{j}$. Find their distance apart at the moment when the velocity of P relative to Q is in the direction QP.

14   A particle starts with a velocity of $-2\mathbf{i}$ and moves with an acceleration of $8\cos 2t\,\mathbf{i} + 4\mathbf{j}$. Find the magnitude of the acceleration when it is first moving parallel to the y axis.

15   A particle P starts at the point $(4,0)$ and moves anticlockwise in a circle, centre the origin, with an angular speed of $1\ \mathrm{rad\,s^{-1}}$. Another particle Q starts at the same time at $(2,0)$, and moves clockwise in a concentric circle with the same angular speed. Prove that the velocity of P relative to Q is $-2\sin t\,\mathbf{i} + 6\cos t\,\mathbf{j}$, and hence show that the relative speed of the two particles has minimum and maximum values of 2 and 6, respectively.

17

Simple Harmonic Motion

In this final chapter we examine a type of motion which occurs very frequently both in nature and in scientific laboratories. This is an *oscillatory* motion called *Simple Harmonic Motion* (SHM), which may be defined as follows.

> *A particle is said to move with SHM when its acceleration along its path is directed towards a certain fixed point in this path, and is directly proportional to the particle's distance (measured along the path) from the fixed point.*

(The motion is usually, but not always, in a straight line.)

Since the acceleration is always towards the fixed point, the particle decelerates as it moves away from the point and eventually comes to instantaneous rest. It then accelerates as it moves towards the point, attaining its maximum speed when it reaches it and then decelerating once more as it moves away from the point in the opposite direction. The motion is therefore oscillatory, the fixed point being the centre of the oscillations. Since the magnitude of the acceleration is proportional to the particle's distance from the fixed point, the maximum acceleration occurs at the two extreme positions, and the acceleration is zero when the particle is at the fixed point. These important features of an SHM are shown in Fig. 17.1, in which the motion is considered to take place along an x-axis, the origin O being the fixed point and A and B being the two extreme positions.

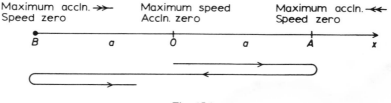

Fig. 17.1

The distance OA (which is equal to OB owing to the symmetry of the motion) is called the *amplitude* of the SHM and denoted by a. The symbol a consequently should not be used for acceleration in work on SHM; the latter quantity can be denoted by d^2x/dt^2 or $\ddot{x}$. (A dot over a symbol indicates differentiation with respect to time.)

Standard equations for Simple Harmonic Motion

The definition is normally expressed mathematically as follows.

$$\frac{d^2x}{dt^2} = -\omega^2 x.$$

Note: (a) that the minus sign is required because the acceleration d^2x/dt^2 is negative when the displacement x is positive, and vice versa; (b) that the constant of proportionality is denoted by ω^2 because its square root occurs in the equations derived from the definition.

We saw in the last chapter that in order to integrate an equation in which acceleration is expressed as a function of displacement, we must denote the acceleration by $v\,dv/ds$ or $v\,dv/dx$. To obtain further standard equations we therefore proceed as follows:

$$v\frac{dv}{dx} = -\omega^2 x$$

$$\therefore \ v\,dv = -\omega^2 x\,dx$$

$$\therefore \ \frac{v^2}{2} = -\frac{\omega^2 x^2}{2} + c.$$

Since $v = 0$ when $x = a$, $c = \omega^2 a^2/2$; hence

$$\frac{v^2}{2} = -\frac{\omega^2 x^2}{2} + \frac{\omega^2 a^2}{2},$$

from which

$$v^2 = \omega^2(a^2 - x^2).$$

This is the first standard SHM equation, other than the definition.

To obtain an equation for x, we write $v = dx/dt$ and integrate again. From the equation just obtained we have

$$\frac{dx}{dt} = \pm\omega\sqrt{a^2 - x^2}$$

$$\therefore \ \frac{dx}{\sqrt{a^2 - x^2}} = \pm\omega\,dt$$

The LHS can now be integrated by means of the standard substitution $x = a\sin\theta$, giving

$$\text{arc}\sin(x/a) = \pm\omega t + k$$

$$\therefore \ x = a\sin(\pm\omega t + k).$$

The $\pm$ in this equation can in fact be omitted since the arbitrary constant k allows for the minus on account of the trigonometric relationship $\sin(-\theta + \pi) = \sin\theta$. The value of k itself depends upon the point at which the SHM is considered to start. Now it is generally convenient either to consider the motion to start at the moment when the particle is at O (and moving in the positive direction), or to take the extreme point A as the starting position. Each of these procedures gives a simple equation for x in terms of t.

Case (a) $t = 0$ when $x = 0$ and the velocity is positive

Substitution of these values into the above equation gives $k = 0$ and we thus have

$$x = a \sin \omega t.$$

Case (b) $t = 0$ when $x = a$

Substitution of these values gives
$$a = a \sin k,$$
from which
$$k = \pi/2.$$

Since $\sin(\theta + \pi/2) = \cos \theta$, we thus have

$$x = a \cos \omega t.$$

Note: It should be remembered, when this equation is used, that although t is measured from the extreme position, x still denotes the distance from the central point.

Periodic time, T

The periodic time (or simply *period*) of an SHM is defined to be the time of *one complete oscillation*, e.g. the time the particle takes for the journey $A \rightarrow B \rightarrow A$ (see Fig. 17.1), or the journey $O \rightarrow A \rightarrow B \rightarrow O$. An expression for the value of T can be obtained from the equation $x = a \sin \omega t$. The value of x passes through one complete cycle whenever ωt increases by 2π, and thus when t increases by $2\pi/\omega$. This is consequently the value of T. A quantity related to T which is sometimes used in SHM work is the *frequency* of the oscillations, denoted by f. This is defined to be the number of oscillations per unit time and its value is $1/T$ or $\omega/2\pi$.

The set of standard SHM equations which should be learned by heart is as follows.

$$\ddot{x} = -\omega^2 x$$
$$v^2 = \omega^2(a^2 - x^2)$$
$$x = a \sin \omega t \qquad (t = 0 \text{ when } x = 0)$$
$$x = a \cos \omega t \qquad (t = 0 \text{ when } x = a)$$
$$T = 2\pi/\omega$$

The following points should also be noted.
(a)   The above set contains an equation for v in terms of x, but not one giving v in terms of t. Such an equation can immediately be obtained, however, simply by differentiating one of the equations for x:

If $t = 0$ when $x = 0$, $v = a\omega \cos \omega t$;

if $t = 0$ when $x = a$, $v = -a\omega \sin \omega t$.

(b) The maximum velocity occurs when $x = 0$, and from the equation $v^2 = \omega^2(a^2 - x^2)$ we thus have

$$(v_{max})^2 = \omega^2 a^2$$

and hence

$$v_{max} = \pm \omega a.$$

(c) The maximum acceleration occurs when $x = \pm a$, and its value is $\pm \omega^2 a$.

Worked examples

Example 1 A particle moving with SHM performs 15 complete oscillations, of amplitude 3 m, every minute. Find (a) the maximum acceleration, (b) the maximum speed, (c) the average speed, (d) the speed when the particle is 2 m from the central position, (e) the particle's position 0.4 s after leaving the central position.

The period T is clearly $\frac{60}{15} = 4$ s. Hence since $T = 2\pi/\omega$, we have $\omega = 2\pi/T = \pi/2$.

(a) The value of the maximum acceleration is $\omega^2 a$ (see above). Hence

$$\text{maximum acceleration} = \frac{\pi^2}{4} \times 3$$

$$= 7.402 \text{ m s}^{-1}.$$

(b)
$$v_{max} = \pm \omega a \quad \text{(standard result – see above)}$$
$$\therefore \text{ maximum speed} = 3\pi/2$$
$$= 4.712 \text{ m s}^{-1}.$$

(c)
$$\text{Average speed} = \frac{\text{distance covered in one oscillation}}{\text{time taken}}$$

$$= \frac{4a}{T}$$

$$= 3 \text{ m s}^{-1}.$$

(d) Using $v^2 = \omega^2(a^2 - x^2)$ and substituting $x = 2$, $a = 3$, $\omega = \pi/2$, we have

$$v^2 = \frac{\pi^2}{4}(9 - 4)$$

from which

$$\text{speed} = 3.512 \text{ m s}^{-1}.$$

(e) Using $x = a \sin \omega t$ and substituting $t = 0.4$, we have

$$x = 3 \sin(\pi/2 \times 0.4)$$

$$= 1.763 \text{ m}.$$

Example 2 A particle moving with SHM has speeds of 4 m s^{-1} and 2 m s^{-1} when its distances from the central point are respectively 30 cm and 40 cm. Find the amplitude.

Using

$$v^2 = \omega^2(a^2 - x^2)$$

we have
$$16 = \omega^2(a^2 - 0.09) \quad (1)$$
and
$$4 = \omega^2(a^2 - 0.16). \quad (2)$$

We can now eliminate ω by dividing (1) by (2):

$$4 = \frac{a^2 - 0.09}{a^2 - 0.16}$$

$$\therefore \quad 4a^2 - 0.64 = a^2 - 0.09$$
$$\therefore \quad 3a^2 = 0.55$$

from which
$$a = 0.4282 \text{ m} \quad \text{or} \quad \textbf{42.82 cm}.$$

Example 3   A particle starts at rest and moves with SHM, the period being T and the amplitude a. For the instant at which the particle has been moving for two-thirds of the time it takes to return to rest, find (a) the distance it has travelled, (b) its speed.

Since the particle starts at rest, we have $t = 0$ when $x = a$. The equation relating x and t is thus
$$x = a \cos \omega t.$$

(a) The particle first comes to rest again when it has completed half an oscillation, and thus after a time of $T/2$. We therefore require the value of x when $t = T/2 \times 2/3 = T/3$. Since $\omega = 2\pi/T$, substitution into the above equation gives

$$x = a \cos \frac{2\pi}{T} \frac{T}{3}$$

$$= -a/2.$$

This is the displacement of the particle from the *central* position after a time of $T/3$; hence we have the situation shown in Fig. 17.2. The distance travelled is therefore **3a/2**.

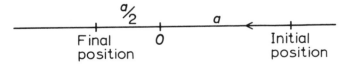

Fig. 17.2

(b) Differentiating the above equation for x, we have

$$v = \frac{dx}{dt} = -a\omega \sin \omega t.$$

Substituting $\omega = 2\pi/T$ and $t = T/3$, we obtain

$$v = -a\frac{2\pi}{T} \sin \frac{2\pi}{3}$$

$$= -\frac{2\pi a}{T} \frac{\sqrt{3}}{2}.$$

The particle's speed is thus $\dfrac{\pi a \sqrt{3}}{T}$.

Example 4 At a certain moment a particle describing SHM is at a point P, and moving away from its central position with a speed of 2 m s^{-1}. After 3 s it returns to P for the first time, and 4.5 s after this it comes to instantaneous rest. Find the period and amplitude of the motion.

By the symmetry of the motion, since the particle takes 3 s to return to P, it must take 1.5 s to come to instantaneous rest for the first time, as shown in Fig. 17.3.

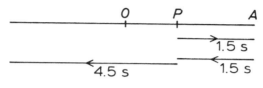

Fig. 17.3

Clearly $T = 2 \times 6 = \textbf{12 s}$, and hence $\omega = 2\pi/12 = \pi/6$. Now using

$$x = a \cos \omega t,$$

we have

$$v = -a\omega \sin \omega t.$$

Since the particle is moving at 2 m s^{-1} to the left when the time measured from A is 1.5 s, we have $v = -2$ when $t = 1.5$ and hence

$$2 = \frac{a\pi}{6} \sin \frac{\pi \times 1.5}{6},$$

from which

$$a = \textbf{5.402 m}.$$

Exercise 17a

1 The period and amplitude of an SHM are 2 s and 0.5 m, respectively. Find the maximum speed, the average speed and the maximum acceleration.

2 A particle describing SHM performs 10 complete oscillations, of amplitude 2 m, each minute. Find (a) its average speed, (b) its speed when 80 cm from its central point, (c) its distance from the centre 0.5 s after leaving this point.

3 A particle oscillates with SHM between two points 2.6 m apart, taking 4 s to pass from one point to the other. Find (a) its acceleration at the two extreme points, (b) its speed when midway between them.

4 A particle moving with SHM has speeds of 2 m s^{-1} and 1 m s^{-1} when distant 80 cm and 120 cm from its central position. Find the amplitude and period.

5 A particle moving with SHM is 60 cm from its central position $\frac{1}{4}$ s after leaving it. If the amplitude is 1.2 m, what is the period?

6 The maximum speeds and accelerations of an SHM are respectively 3 m s^{-1} and 4.5 m s^{-2}. Find the amplitude and period.

7 A particle performs 30 simple harmonic oscillations per minute, at an average speed of 3 m s^{-1}. Find the amplitude, and the particle's speed when 50 cm from its position of instantaneous rest.

8   A particle moves with an SHM of amplitude 2 m and period 8 s. What is its speed 1 s after leaving its position of instantaneous rest?

9   A particle performs 40 simple harmonic oscillations per minute and has a speed of 4 m s^{-1} when 2 m from its central position. Find the amplitude, and the particle's speed when 1 m from the central position.

10   A particle performs simple harmonic oscillations of amplitude 1.4 m. If its speed when 80 cm from the central point is 2 m s^{-1}, what is its acceleration at this point?

11   A particle moving with SHM has an acceleration of 2 m s^{-2} when 50 cm from the central point. If its maximum acceleration is 3 m s^{-2}, what is its maximum speed?

12   A particle performs 90 simple harmonic oscillations per minute, and has a speed of 2 m s^{-1} when 60 cm from the central position. Finds its average speed.

13   A particle starts at rest and moves with SHM. After 1 s its speed is 50 cm s^{-1}, and after another 5 s it is at rest again. What distance has it travelled?

14   A particle moving with SHM is 2 m from its central point O, and moving towards O. After an interval of 1.8 s it reaches for the first time the other point in its path which is 2 m from O, and 0.2 s after this it is at rest. What is the amplitude of the motion?

15   A particle moving with SHM is at a point P, 50 cm from its central point, and moving away from the central point. 1.5 s later it returns to P for the first time, and after another 4.5 s it reaches P for the second time. Find the amplitude.

16   A particle starts at rest and moves with SHM. After 1 s it has travelled 2 m and after travelling another 8 m it is at rest again. What is its maximum speed?

17   For what percentage of its period does a body moving with SHM have less than three-quarters of its maximum speed?

18   A particle performs simple harmonic oscillations of period T and amplitude a. Obtain expressions for the distance from the centre and the speed at a time of $T/6$ after the particle is at its central point.

19   A particle moving with SHM of amplitude a has a maximum speed of V. Find its distance from the central point when its speed is $0.8\ V$.

20   A particle performs simple harmonic oscillations of period T. Find the time taken for the speed to drop from its maximum value to half that value.

21   What is the greatest possible amplitude of an SHM of frequency f, if the speed must not exceed V?

Practical examples of Simple Harmonic Motion

Many oscillatory motions are at least approximately simple harmonic, and the procedure for demonstrating that any particular motion is of this kind can be conveniently divided into the following steps.

1.   Locate the *equilibrium position* of the oscillating body – the position in which there is no resultant force on it. Since the acceleration of the body is then zero, the equilibrium position must be the central point of the oscillation.

2.   Consider any other position occupied by the body during its motion, and let its displacement from the equilibrium position then be x.

3.   Obtain an expression in terms of x for the resultant force on the body, and by using $F = ma$, deduce an expression for its acceleration. This should have the

form $-\omega^2 x$, and it will thus have been shown that the motion obeys an equation of the form $d^2x/dt^2 = -\omega^2 x$. Since this is the defining equation of an SHM, it will follow that the motion is simple harmonic.

The value of ω depends on the physical properties of the oscillating system, and for any particular kind of system – such as the simple pendulum – ω is given by a standard expression whch can be learned by heart. Alternatively, the expression for the period – given by $2\pi/\omega$ – can be learned. The other constant needed to specify an SHM, its amplitude, varies according to the manner in which the oscillations are started. It should be noted that for a perfect SHM, the values of ω and T are independent of the amplitude.

The procedure outlined above will now be applied to the two simplest practical examples, namely the simple pendulum and the elastic spring or string.

The simple pendulum

This consists of a heavy particle which swings in a vertical plane at the end of a light string whose other end is attached to a fixed point. We shall show that when the amplitude of the oscillations is small, they are approximately simple harmonic.

Let the length of the string be l and the mass of the particle m. The equilibrium position of the particle is clearly its lowest point. Consider any other position of the particle during its swing, and let the distance from the lowest point, measured along the path, be x. Let the angular displacement of the string from the vertical be θ (Fig. 17.4).

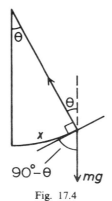

Fig. 17.4

The resultant force on the particle is in the direction of the tangent to its path, and is given by $mg \sin \theta$. Using $F = ma$, we thus have

$$\frac{d^2x}{dt^2} = -g \sin \theta$$

$$= -g\theta \quad \text{approximately if } \theta \text{ is small.}$$

Since $x = l\theta$, $\theta = x/l$ and hence

$$\frac{d^2x}{dt^2} = -\frac{g}{l}x.$$

This equation is of the form $d^2x/dt^2 = -\omega^2 x$, and the motion is therefore simple harmonic with $\omega^2 = g/l$. It follows that the period is given by

$$T = 2\pi \sqrt{\frac{l}{g}}.$$

Particle attached to spring (or elastic string)

Consider a particle of mass m which oscillates at the end of a spring of natural length l and modulus λ, whose other end is attached to a fixed point. We shall show that both horizontal and vertical oscillations of the particle are simple harmonic, with the same period.

Case (a): spring horizontal   Let the oscillations take place on a smooth horizontal plane, so that the equilibrium position is that at which the spring has its natural length. Consider any other position of the particle, in which its displacement from the equilibrium position is x (Fig. 17.5).

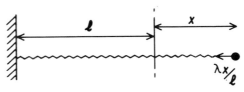

Fig. 17.5

The resultant force on the particle is simply the tension in the spring, which, by Hooke's law, is $\lambda x/l$. Using $F = ma$, we thus have

$$\frac{d^2x}{dt^2} = -\frac{\lambda x}{ml}.$$

This equation is of the form $d^2x/dt^2 = -\omega^2 x$, and the motion is therefore simple harmonic with $\omega^2 = \lambda/ml$. It follows that the period is given by

$$T = 2\pi \sqrt{\frac{ml}{\lambda}}.$$

It should be noted that when a particle moves horizontally at the end of an elastic *string*, complete oscillations are impossible. The motion is however simple harmonic while the string is taut.

Case (b): spring vertical   In this case we begin by locating the equilibrium position. Let the extension in this position be e, and consider any other position of the particle, in which the total extension is $e + x$ (Fig. 17.6).

The resultant upward force, when the extension is $e + x$, is

$$\frac{\lambda(e + x)}{l} - mg$$

$$= \frac{\lambda e}{l} + \frac{\lambda x}{l} - mg.$$

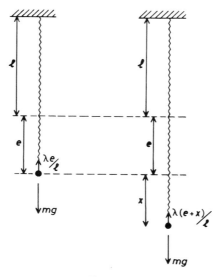

Fig. 17.6

From the first diagram, however, in which the particle is in equilibrium, we have $\lambda e/l = mg$. It follows that the resultant upward force is $\lambda x/l$, and, as in case (a),

$$\frac{d^2x}{dt^2} = -\frac{\lambda x}{ml}.$$

The motion is therefore simple harmonic with a period again given by

$$T = 2\pi \sqrt{\frac{ml}{\lambda}}.$$

It is possible in this case for complete oscillations to occur when the spring is replaced by an elastic string. The condition is that the string always remains taut, and thus that the amplitude is less than the value of e.

Worked examples

Example 1 Find the length of a simple pendulum which takes $\frac{3}{4}$ s for each swing between rest positions. If the maximum speed of the bob is 50 cm s^{-1}, what is the angle through which the string turns at each swing?

Substituting $T = 1.5$ into the equation $T = 2\pi\sqrt{(l/g)}$, we have

$$1.5 = 2\pi \sqrt{\frac{l}{10}}$$

from which

$$l = 0.5699 \text{ m}.$$

The maximum speed of 0.5 m s^{-1} is given by the expression ωa, and $\omega = 2\pi/T = 2\pi/1.5$. Hence

$$\frac{2\pi a}{1.5} = 0.5,$$

from which
$$a = 0.1194 \text{ m}.$$

Now consider Fig. 17.7, which shows the relationship between the amplitude a and the required angle θ.

0.5699 m

0.1194 m

Fig. 17.7

We have

$$\theta = \frac{2 \times 0.1194}{0.5699}$$

$$= 0.419 \text{ rad} = \mathbf{24°} \text{ to the nearest degree.}$$

It should be noted that the second part of this question could have been done by using the principle of conservation of energy (see Chapter 5). Energy equations can often be used as an alternative to SHM methods, though these equations in general relate velocities and distances, and do not provide information about *time*. If the latter quantity is involved in a problem, therefore, the SHM equations must be used.

Example 2   A pan of mass 1 kg is suspended by a vertical spring of natural length 50 cm and modulus 50 N. A 2 kg particle is dropped onto the pan, striking it at a speed of 3 m s^{-1}, and the two then move together. Find the period and amplitude of the subsequent oscillations, and the time between the impact and the moment at which the pan first reaches its lowest point.

The period is given by the standard equation

$$T = 2\pi \sqrt{\frac{ml}{\lambda}}.$$

Substituting $m = 3$ (since the oscillating mass is that of particle and pan combined), $l = 0.5$ and $\lambda = 50$, we have

$$T = 2\pi \sqrt{\frac{3}{100}} = 1.088 \text{ s}.$$

It should also be noted that $\omega = \sqrt{(100/3)}$; this will be needed for the other parts of the question.

We begin the calculation of the amplitude by locating the two equilibrium positions; that of the pan alone, and that of the pan and particle combined. For the pan alone we have

$$mg = \frac{\lambda e}{l}$$

that is,

$$10 = \frac{50e}{0.5}.$$

from which $e = 0.1$ m.

For the pan and particle combined we have

$$30 = \frac{50e}{0.5},$$

from which $e = 0.3$ m.

The second equilibrium position is the centre of the SHM; hence the value of x at the start of the motion is $0.3 - 0.1 = 0.2$ m.

Next we find the initial velocity, v. Using the principle of conservation of momentum, we have

$$2 \times 3 = 3v$$

and hence

$$v = 2 \text{ m s}^{-1}.$$

Substituting now into the equation $v^2 = \omega^2(a^2 - x^2)$, we obtain

$$4 = \frac{100}{3}(a^2 - 0.04)$$

from which $a = \mathbf{0.4 \text{ m}}$.

The time the system takes to come to rest is the time it takes to reach its new equilibrium position (the centre of the oscillations) plus one quarter of the period. Letting t be the first of these and using the equation $x = a \sin \omega t$, we have

$$0.2 = 0.4 \sin t \sqrt{\frac{100}{3}},$$

from which $t = 0.09069$ s. Since a quarter of the period is $(\pi/2)\sqrt{0.03} = 0.2721$, the total time is $\mathbf{0.3628 \text{ s}}$.

Example 3 An elastic string of natural length l and modulus $2mg$ is attached at one end to a fixed point and at the other to a particle of mass m which hangs vertically. The particle is pulled down a further distance l and released. Obtain an expression for the time it takes to reach its highest point.

We begin by locating the equilibrium position. Letting the extension in this position be e, we have

$$\frac{\lambda e}{l} = mg,$$

that is,

$$\frac{2mge}{l} = mg,$$

and hence

$$e = \frac{l}{2}.$$

Since the particle is pulled down a distance greater than $l/2$, the string will become slack before the particle reaches its highest point, and the motion is therefore simple harmonic for only part of the upward journey (Fig. 17.8). The particle starts at A and oscillates about B, the equilibrium position. At C the string becomes slack, and the motion ceases to be simple harmonic.

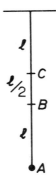

Fig. 17.8

Stage 1: AB   The time for this stage is one quarter of the period, i.e.

$$\frac{1}{4} 2\pi \sqrt{\frac{ml}{\lambda}} = \frac{\pi}{2} \sqrt{\frac{l}{2g}}.$$

Stage 2: BC   Using $x = a \sin \omega t$, we have

$$\frac{1}{2} = l \sin t \sqrt{\frac{2g}{l}}$$

$$\therefore \ t \sqrt{\frac{2g}{l}} = \frac{\pi}{6}$$

and

$$t = \frac{\pi}{6} \sqrt{\frac{l}{2g}}.$$

Stage 3: C to the highest point   During this stage the particle moves under its own weight alone, and thus with a constant downward acceleration of g. We begin by finding the speed at C. Using

$$v^2 = \omega^2 (a^2 - x^2),$$

we have

$$v^2 = \frac{2g}{l} \left(l^2 - \frac{l^2}{4} \right) = \frac{3gl}{2}.$$

Hence

$$v = \sqrt{\frac{3gl}{2}}.$$

Now using

$$v = u + at,$$

we have

$$0 = \sqrt{\frac{3gl}{2}} - gt,$$

from which

$$t = \sqrt{\frac{3l}{2g}}.$$

The total time is thus

$$\sqrt{\frac{l}{2g}}\left(\frac{\pi}{2} + \frac{\pi}{6} + \sqrt{3}\right) = \sqrt{\frac{l}{2g}}\left(\frac{2\pi}{3} + \sqrt{3}\right).$$

Exercise 17b

The standard expressions for ω and T are quotable without proof unless proofs are specifically demanded.

1   A spring of natural length 1.5 m and modulus 10 N is attached at one end to a point on a smooth horizontal table, and at the other to a 2 kg particle which rests on the table. The particle is pulled to a distance of 2 m from the fixed point, and released. Prove that the subsequent oscillations are simple harmonic, and find (a) the period, (b) the speed of the particle when the spring is unextended.

2   It is found that a simple pendulum of length 60 cm performs 100 complete oscillations in 156 s. What value of g does this indicate?

3   An elastic string of natural length 50 cm and modulus 40 N is attached at one end to a fixed point and at the other to a 2 kg particle which hangs below the fixed point. Find the extension when the particle is stationary. When the particle is pulled down another 30 cm and released, show that the motion is simple harmonic while the string is extended, and find the particle's speed (a) when it is in the equilibrium position, (b) at the moment when the string becomes slack.

4   A vertical spring supports a body which extends it a distance e when in equilibrium. Show that when the body is pulled down vertically and released, the period of the resulting oscillations is given by $2\pi\sqrt{(e/g)}$. (Here the formula $T = 2\pi\sqrt{(ml/\lambda)}$ can be quoted without proof.)

5   A vertical spring supports a body which extends it 50 cm when in equilibrium. The body is pulled down another 30 cm and released. Find (a) the period of the resulting oscillations, (b) the body's speed after travelling 10 cm, (c) the time it takes to travel this distance.

6   A body attached to the end of a vertical spring makes 2 oscillations every second. Find the extension at the equilibrium position.

7   The bob of a simple pendulum of length 70 cm has a maximum speed of 60 cm s^{-1}. Find the period and amplitude of the oscillations.

8   A spring of natural length 40 cm is doubled in length by a force of 20 N. Find the period of the spring's oscillations when it is carrying a body of mass 250 g.

9   A vertical spring carries a pan of mass 250 g which extends it 10 cm. A 750 g particle is placed on the pan and the system is released. Find (a) the time taken for the pan to return to rest, (b) the maximum acceleration of the pan.

10   A 3 kg particle is attached to one end of a horizontal elastic string of natural length 1 m and modulus 5 N which is fixed at its other end to a point A on a smooth horizontal table. If the particle is pulled to a distance of 1.5 m from A and

released, how long will it take to reach A? (Remember that the motion is in two stages: an SHM followed by a motion with constant speed.)

11   The bob of a simple pendulum of length 50 cm is released when the string is at $10°$ to the vertical. Find the average speed of the bob during the resulting oscillations.

12   A 4 kg particle hangs in equilibrium at the end of a vertical elastic string of natural length 90 cm and modulus 60 N. From this position it is projected downwards with a speed of $2 \, \mathrm{m \, s^{-1}}$. Find its speed 0.5 s later.

13   A simple pendulum of length l is held at a small angle θ to the vertical and released. Prove that the time it takes to turn through an angle of $3\theta/2$ is $\frac{2}{3}\pi\sqrt{(l/g)}$.

14   A particle hangs in equilibrium at the end of a vertical elastic string, causing an extension of e. It is pulled down a further distance e and released. Obtain an expression for its speed when it has travelled a distance of $e/2$.

15   A vertical spring carries a pan of mass m which extends it a distance e. A particle of mass $4m$ is placed on the pan and released. Obtain expressions for the maximum speed and acceleration of the pan in the subsequent motion.

16   A simple pendulum of length 80 cm is held at an angle of $15°$ to the vertical and released. What is the speed of the bob after 0.3 s?

17   A pan of mass 500 g hangs in equilibrium at the end of a vertical elastic string of natural length 120 cm and modulus 20 N. A $1\frac{1}{2}$ kg particle is placed on the pan and released. Find (a) the maximum speed attained by the pan, (b) the maximum tension in the string, (c) the time the pan takes to travel 50 cm.

18   An elastic string of natural length $2l$ and modulus λ is stretched between two fixed points on a smooth horizontal table and a particle of mass m is attached to its mid-point. Show that if the particle is slightly displaced in the direction of the string – the string remaining taut – the resulting oscillations are simple harmonic with a period of $2\pi\sqrt{(ml/2\lambda)}$.

19   An elastic string of natural length 50 cm and modulus 10 N is stretched between two points 1 m apart on a smooth horizontal table. A 2 kg particle is attached to the mid-point, pulled in the direction of the string to a point at which one of the two sections just becomes slack, and released. Find the time it takes (a) to return to its equilibrium position, (b) to travel 40 cm.

20   A 500 g particle hangs in equilibrium at the end of a vertical elastic string of natural length 1 m and modulus 20 N. The particle is pulled down a further distance of 50 cm, and released. Find the time it takes to reach its highest point.

21   A particle of mass m hangs in equilibrium at the end of a vertical elastic string of natural length 1 m whose upper end is attached to a fixed point P. The particle is pulled down to a distance 2 m below P, and released. Given that it just reaches P, show by an energy method that the modulus of the string is $4mg$, and use this to find the time the particle takes to reach P.

22   A particle attached to the lower end of a vertical spring performs 60 oscillations, of amplitude 20 cm, every minute. Find its maximum speed. Given that when it has this speed the particle picks up another particle of equal mass, previously at rest, find to the nearest cm the amplitude of the new oscillations which occur.

Revision Exercises

The revision exercises correspond with the chapters in the material they cover. Thus Exercise R1 deals with the topics in Chapter 1, and so on.

Exercise R1

1   A uniformly accelerating body starts with a velocity of $36\,\text{km}\,\text{h}^{-1}$, and $3\,\text{s}$ later has travelled $48\,\text{m}$. Find its acceleration and final velocity.

2   A body thrown vertically upwards from the ground has an upward velocity of $20\,\text{m}\,\text{s}^{-1}$ after $\frac{1}{2}\,\text{s}$. After what further time does it reach its greatest height, and what is that height?

3   A stone is thrown vertically upwards at $5\,\text{m}\,\text{s}^{-1}$ from the top of a building $60\,\text{m}$ high. Find the time it takes to reach the ground.

4   A particle starts at a point P with a speed of $20\,\text{m}\,\text{s}^{-1}$, and after moving with a uniform deceleration for $4\,\text{s}$ has a speed of $12\,\text{m}\,\text{s}^{-1}$ in the opposite direction. How far is it then from P, and after what further time does it return to P?

5   After travelling for $6\,\text{s}$, a uniformly accelerating body has covered $45\,\text{m}$ and has a velocity of $54\,\text{km}\,\text{h}^{-1}$. Find its acceleration.

6 A uniformly accelerating body travels $10\,\text{m}$ in the first 2 seconds and $8\,\text{m}$ in the next second. Find its acceleration and initial velocity.

7   A ball thrown vertically upwards is more than $4\,\text{m}$ above its starting point for $1.6\,\text{s}$. Find the total time between its being thrown and returning to its starting point.

8   A uniformly decelerating train of length $30\,\text{m}$ takes $4\,\text{s}$ to pass completely through a station of length $32\,\text{m}$. Given that the front of the train leaves the station at $16\,\text{m}\,\text{s}^{-1}$, find the deceleration.

9   A body starts from rest, accelerates uniformly to a speed of $40\,\text{m}\,\text{s}^{-1}$, travels at this speed for a time, then decelerates uniformly to rest. If the distances covered in the three stages are in the ratio $1:2:2$, and the total time is $80\,\text{s}$, find the initial acceleration.

10   Bodies A and B start together and move along the same straight line. A moves at $36\,\text{m}\,\text{s}^{-1}$ for a time t, then decelerates at $2\,\text{m}\,\text{s}^{-2}$, while B starts at rest and accelerates at $4\,\text{m}\,\text{s}^{-2}$. Find t, given that the bodies are together again after a time of $2t$.

11   Bodies A and B start a distance d apart and move towards each other. A starts with a speed of u and decelerates at a, while B starts at rest and accelerates at $2a$. Given that their speeds are equal when they meet, prove that $7u^2 = 18ad$.

12   A body starts with a speed of u, accelerates uniformly to twice this speed, travels at constant speed for the same time, then decelerates uniformly to rest, travelling the same distance in the third stage as in the first. Prove that the average speed for the whole journey is $10u/7$.

Exercise R2

1   A force of 90 N acts vertically upwards on a body. Find (a) the acceleration if the weight of the body is 50 N, (b) the body's mass if its upward acceleration is $5\,\mathrm{m\,s^{-2}}$.

2   A body of weight 400 N accelerates downwards at $3\,\mathrm{m\,s^{-2}}$. What upward force is acting on it?

3   A 750 g body starts at rest and is pushed 24 m in 3 s along a rough horizontal surface by a force of 20 N. Find the frictional force on the body.

4   A force of $9\,W$ acts vertically upwards on a body of weight $4\,W$. If the body accelerates at $3g/4$, what retarding force is acting?

5   A force of 50 N pushes a 12 kg body, initially at rest, along a rough horizontal surface for 4 s. The force is then removed and the body comes to rest in another 6 s. Given that a constant frictional force acts throughout the motion, find this force.

6   A force of $7x$, acting vertically upwards on a certain body, causes twice the upward acceleration caused by an upward force of $5x$. Find the mass of the body in terms of x and g.

7   A man of weight W stands on a weighing machine which accelerates (a) upwards at $0.25\,g$, (b) downwards at $0.8\,g$. What weights are recorded?

8   A box of mass $1\frac{1}{4}$ kg carrying a block of mass $2\frac{1}{2}$ kg is lifted vertically upwards by a force of 60 N. Find (a) the acceleration, (b) the force between the box and the block.

9   A block of weight 100 N stands on a block of weight 140 N, and an upward force of 300 N is applied to the latter. Find the force between the two blocks.

10   Bodies of mass 4 kg, 6 kg and 10 kg are connected by strings and pulled along a smooth horizontal surface by a force of 60 N applied to the 4 kg body. Find the tensions in the strings.

11   Bodies of mass $8m$, $5m$ and $12m$ are connected by strings and pulled along a smooth horizontal table by a force of $100\,mg$ applied to the body of mass $12m$. Find the tensions in the strings in terms of m and g.

12   A pan of mass 25 g carrying a body of mass 175 g is lifted vertically upwards with an acceleration of $6\,\mathrm{m\,s^{-2}}$. Find (a) the external force applied to the pan, (b) the reaction between the pan and the body.

13   Bodies of mass 2 kg, 3 kg and 7 kg are connected by strings and pulled along a smooth horizontal table by a force applied to the 7 kg body. Given that the tension in the string between the 3 kg and 7 kg bodies is 25 N, find (a) the acceleration, (b) the applied force, (c) the tension in the other string.

14   A pile of blocks with weights (from the bottom upwards) of $7W$, $5W$, $2W$ and W, is lifted by a force applied to the bottom block. Given that the force between the bottom two blocks is $24W$, find (in terms of W and g) (a) the acceleration, (b) the applied force, (c) the force between the middle two blocks.

15   Bodies A, B, C, with masses of 1 kg, 3 kg and 4 kg, are connected by strings and pulled along a smooth horizontal surface by a force of $3F$ applied to C. A retarding force of F, directly opposite to the force of $3F$, acts on A. Given that the tension in the string between B and C is 24 N, find the tension in the string between A and B.

Exercise R3

When physical quantities are given in terms of **i** and **j**, the units are SI.

1 $ABCD$ is a quadrilateral. Express as single vectors (a) $\overrightarrow{CA} + \overrightarrow{DC}$, (b) $\overrightarrow{BC} - \overrightarrow{BA}$, (c) $\overrightarrow{BD} + \overrightarrow{CA} + \overrightarrow{DC}$, (d) $\overrightarrow{AD} - \overrightarrow{CD} - \overrightarrow{AB}$.

2 O is the origin, and P, Q, R are the points $(-6, 2)$, $(3, -7)$, $(-5, -3)$, respectively. Express the following vectors in terms of **i** and **j**: (a) $\overrightarrow{PO}$, (b) $\overrightarrow{QR}$, (c) $\overrightarrow{RP}$, (d) $\overrightarrow{QS}$, where S is the mid-point of OP, (e) $\overrightarrow{ST}$, where T is the mid-point of QR.

3 Find the magnitudes and inclinations to the positive x-axis of the following vectors: (a) $7\mathbf{i} + 24\mathbf{j}$, (b) $-\mathbf{i} + 2\mathbf{j}$, (c) $12\mathbf{i} - 16\mathbf{j}$, (d) $-12\mathbf{i} - 4\mathbf{j}$.

4 A, B, C, D have position vectors $5\mathbf{i} + \mathbf{j}$, $2\mathbf{i} - 9\mathbf{j}$, $-4\mathbf{i} + 2\mathbf{j}$, $-\mathbf{i} + 12\mathbf{j}$, respectively. Express $\overrightarrow{AB}$ and $\overrightarrow{DC}$ in terms of **i** and **j**. What kind of figure is $ABCD$?

5 Express in terms of **i** and **j** (a) a vector of magnitude 100 in the direction of the vector $24\mathbf{i} + 7\mathbf{j}$, (b) a vector of magnitude 39 in the direction of $5\mathbf{i} - 12\mathbf{j}$, (c) a unit vector in the direction of $-9\mathbf{i} + 12\mathbf{j}$, (d) a vector of magnitude 2 at $+60°$ to the positive x-axis.

6 Find (a) the two vectors of magnitude 5 which are perpendicular to the vector $4\mathbf{i} + 3\mathbf{j}$, (b) the two vectors of magnitude 85 which are perpendicular to the vector $8\mathbf{i} - 15\mathbf{j}$.

7 A 10 kg particle accelerates at $3 \, \text{m s}^{-2}$ in the direction of the vector $8\mathbf{i} - 6\mathbf{j}$. Express the force on the particle in terms of **i** and **j**.

8 A force of magnitude $4\sqrt{5}$ acts on a 500 g particle in the direction of the vector $2\mathbf{i} - \mathbf{j}$. Express the particle's acceleration in terms of **i** and **j**.

9 A 6 kg particle starts at rest, and, owing to a constant force, has a velocity of $12\mathbf{i} - 7\mathbf{j}$ after 2 s. Express the force on it in terms of **i** and **j**.

10 Forces of 50 N and 51 N act in the directions of the vectors $7\mathbf{i} + 24\mathbf{j}$ and $-8\mathbf{i} - 15\mathbf{j}$, respectively. Express the resultant force in terms of **i** and **j**.

11 Forces of $5\mathbf{i} - 9\mathbf{j}$ and $-2\mathbf{i} + 5\mathbf{j}$ acts on a 500 g particle which is initially at rest at the origin. Find the position and velocity of the particle after 3 s.

12 Two forces act on a 3 kg particle, causing it to accelerate in the direction of the negative x-axis at $5 \, \text{m s}^{-2}$. If one of the forces is $-20\mathbf{i} + 9\mathbf{j}$, what is the other?

13 Three forces act on a 2 kg particle, causing it to accelerate in the direction of the vector $3\mathbf{i} + \mathbf{j}$ at $6\sqrt{10} \, \text{m s}^{-2}$. If two of the forces are $9\mathbf{i} + 2\mathbf{j}$ and $15\mathbf{i} - 7\mathbf{j}$, what is the third force?

14 Forces of $25\mathbf{i} + 70\mathbf{j}$, $a\mathbf{i} + 35\mathbf{j}$ and $-12\mathbf{i} + b\mathbf{j}$ act on a 50 kg particle, causing it to move from the origin, where it is initially at rest, to the point $(14, 48)$ in 5 s. Find a and b.

Exercise R4

1 Use the parallelogram method to calculate the magnitude of the resultants of the following pairs of forces: (a) 7 N and 9 N, inclined at $90°$, (b) 3 N and 5 N, inclined at $40°$, (c) 10 N and 24 N, inclined at $90°$, (d) 15 N and 23 N, inclined at $130°$.

2   Forces of 4 N and 7 N act away from a certain point and have a resultant of 9 N. Find the angle between the two forces.

3   Forces of 12 N and 16 N act towards a certain point and have a resultant of 11 N. Find the angle between the two forces.

4   Forces of P and $P\sqrt{3}$ are inclined at an angle of 30°. Prove that the magnitude of their resultant is $P\sqrt{7}$.

5   Forces of $5P$ and $4P$ have a resultant of $P\sqrt{21}$. Prove that the angle between them is 120°.

6   Find the magnitudes and bearings of the resultants of the following systems of horizontal concurrent forces:

(a) 7 N, 5 N, 4 N, on bearings of 20°, 90°, 240°;

(b) 6 N, 6 N, on bearings of 10°, 120°;

(c) 12 N, 8 N, 20 N, 15 N, on bearings of 65°, 180°, 220°, 305°.

7   Horizontal concurrent forces of P, 8 and Q act on bearings of 60°, 180° and 270°. Find P and Q if the system is in equilibrium.

8   A 6 kg particle is held in equilibrium on a smooth horizontal plane of inclination 40° by a horizontal force. Find this force and the normal reaction of the plane.

9   A particle of weight W is held in equilibrium on a smooth plane of inclination $\tan^{-1} 3/4$ by a string inclined at the same angle to a line of greatest slope. The tension in the string is T and the normal reaction is R.

(a) Find T and R, given that $W = 20$ N.

(b) Find W and R, given that $T = 24$ N.

(c) Prove that $20R = 7W$.

10   Horizontal concurrent forces of 20, P and 15 act on bearings of 60°, 155° and 310°. Given that the resultant R is due east in direction, find P and R.

11   A 2 kg body is subjected to two forces of 10 N, as a result of which it accelerates at 4 m s^{-2}. Find the angle between the two forces.

12   A 5 kg particle is attached by strings of length 70 cm and 240 cm to two points on the same horizontal level which are 250 cm apart. Find the tensions in the strings.

13   A 500 g body lying on the ground is pulled by horizontal forces of 12 N, 9 N and P N acting on bearings of θ, 180° and 270°. Find P and θ if (a) the body remains at rest, (b) it accelerates at 6 m s^{-2} in the due south direction.

14   Horizontal concurrent forces of P, 40 and Q act on bearings of 0°, 120° and 220°. Given that the system is in equilibrium, use Lami's theorem to find P and Q.

15   A string ABC has a 2 kg particle attached to it at B, and A and C are atached to two points on the same horizontal level. Given that $\angle BAC = 35°$ and $\angle BCA = 62°$, find the tensions in AB and BC.

16   A 5 kg ring R is threaded onto a string, the ends of which are attached to two points A and B on the same horizontal level. Owing to a horizontal force on the ring, $\angle RAB = 25°$ and $\angle RBA = 45°$. Find the tension in the string and the horizontal force.

17   A 12 kg block lying in equilibrium on rough horizontal ground is pulled by two ropes, in the same vertical plane, which tend to move the block along the ground in opposite directions. The ropes are inclined at 56° and 24° to the ground. Given that the tension in the first rope is 50 N, and that the ground provides a horizontal frictional force of 40 N, find the tension in the second rope and the normal reaction of the ground.

18 A string *ABCD* carries a particle at *B* and a smooth ring at *C*. The ends *A*, *D* are attached to two points on the same horizontal level, and *AB*, *BC* are inclined to the horizontal at 50° and 20°, respectively, *C* being below the level of *B*. Given that the tension in *AB* is 35 N, find the tension in the rest of the string and the weights of the particle and the ring.

19 A string *ABCD* carries particles at *B* and *C*, the weight of the latter being 25 N. The sections *AB*, *BC* and *CD* are inclined to the horizontal at 40°, 15° and 60°, respectively, *C* being below the level of *B*. Find the tensions in the three sections of string and the weight of the first particle.

Exercise R5

1 A force of 180 N acts vertically upwards on a 12 kg body, increasing its upward speed from $2 \, \text{m s}^{-1}$ to $8 \, \text{m s}^{-1}$. Find (a) the work done by the force, (b) the PE gained by the body.

2 A particle which is suspended by a vertical, non-elastic string of length 50 cm is given a horizontal speed of $3 \, \text{m s}^{-1}$. Find (a) its speed when the string has turned through 60°, (b) the angle the string has turned through when the speed is $2.5 \, \text{m s}^{-1}$.

3 A 2000 kg car has a maximum speed of $30 \, \text{m s}^{-1}$ when driving up a slope of 1 in 20 against a resistance of 1500 N. Find (a) its power, (b) its maximum acceleration at the same speed on the level, if the resistance is unchanged.

4 A 200 tonne train is developing 2700 kW. Find the total retarding force on the train when its speed and acceleration are $12 \, \text{m s}^{-1}$ and $0.8 \, \text{m s}^{-2}$. If, at this speed, the power is reduced by 20%, what will be the new acceleration?

5 Every minute a machine raises 20 blocks, each of mass 50 kg, through a vertical distance of 4 m. The blocks are initially at rest, and their average final speed is $2 \, \text{m s}^{-1}$. Calculate the power the machine is supplying.

6 An elastic string of natural length 60 cm and modulus 120 N is attached at one end to a fixed point *P* on a smooth horizontal table, and at the other to a 500 g particle. The particle is released from a point on the table which is 1 m from *P*. Find (a) its speed after travelling 20 cm, (b) the extension of the string when the speed is $6 \, \text{m s}^{-1}$.

7 A 1400 kg car travels on a level road, against a resistance of 1200 N, at its maximum speed of $40 \, \text{m s}^{-1}$. Given that the resistance is proportional to the square of the speed, find the maximum acceleration of the car when its speed is $20 \, \text{m s}^{-1}$.

8 A 2700 kg vehicle with a 90 kW engine can travel up a slope of inclination arc sin 5/36 at a maximum speed of $15 \, \text{m s}^{-1}$. Find its maximum speed on the level if (a) the resistance is unchanged, (b) the resistance is proportional to the square of the speed.

9 A 40 kg boy travels 30 m along a slide while falling through a vertical distance of 8 m. If he starts at rest and finishes with a speed of $2 \, \text{m s}^{-1}$, what is the average frictional resistance of the slide?

10 A 3 kg particle hangs at the end of a vertical elastic string of natural length 40 cm. The particle is pulled down until the total extension is 20 cm, and released. Given that it just reaches the point at which the top of the string is supported, calculate the modulus of elasticity of the string.

11   Every minute a pump raises 12 tonnes of water, initially at rest, through a vertical distance of 8 m, and delivers it through a pipe of cross sectional area 200 cm². Find the power supplied by the pump.

12   A 600 tonne train can develop 8000 kW. Given that its maximum acceleration on a level track is $0.5\,\mathrm{m\,s^{-2}}$ when its speed is $72\,\mathrm{km\,h^{-1}}$, and that the resistance is proportional to the square of the speed, find (a) its maximum acceleration at a speed of $36\,\mathrm{km\,h^{-1}}$, (b) its maximum speed.

13   A car is driven on a level road with a constant power against a constant resistance. When the speed is v the acceleration is f and when the speed is $2v/3$ the acceleration is $2f$. Find (a) the maximum speed, (b) the speed at which the acceleration is $3f$.

14   A 2500 kg body falls from a height of 3 m onto soft ground which provides an average resistance of 400 kN. Find the distance the body penetrates.

15   An elastic string of natural length 40 cm and modulus 150 N is attached at its upper end to a point P on a smooth plane inclined at $30°$ to the horizontal. The other end is attached to a particle, which is held on the line of greatest slope through P, at a distance of 60 cm from P, and released. Given that it reaches P with a speed of $2\,\mathrm{m\,s^{-1}}$, find the mass of the particle.

16   A car can freewheel down a slope of 1 in N at a steady speed of v, and its maximum speed when driving up the slope is $2v$. The resistance is proportional to the square of the speed. Prove that the car's maximum acceleration when driving down the slope at a speed of $2v$ is $2g/N$.

17   A car can drive up a certain slope at a maximum speed of $3v$. When the car pulls a trailer of equal mass up the slope, the maximum speed is $2v$. Given that the resistance per unit mass is proportional to the square of the speed, prove that either vehicle will freewheel down the slope at a steady speed of $v\sqrt{11}$.

Exercise R6

1 $ABCD$ is a rectangle in which $AB = 12\,\mathrm{cm}$ and $BC = 9\,\mathrm{cm}$. Forces of 6 N, 10 N, 20 N act along BA, BC, DB. Find the resultant moment about A and about the mid-point of CD.

2 $ABCD$ is a line in which $AB = BC = 2\,\mathrm{cm}$ and $CD = 4\,\mathrm{cm}$. Like forces of 10 N, 6 N, 8 N, and an opposite force of 30 N, all perpendicular to the line, act at A, C, D, B, respectively. Find the magnitude of the resultant and its perpendicular distance from A.

3   A uniform rod AB of length 1 m and mass 14 kg rests horizontally on supports distant 10 cm from A and 20 cm from B. Find (a) the forces on the supports, (b) the upward force at A which just causes the rod to tip, (c) the couple which equalises the forces on the supports.

4   Forces of $6\mathbf{j}$, $-8\mathbf{j}$, $F\mathbf{j}$, $4\mathbf{j}$ act at the points $(-4, 0)$, $(1, 0)$, $(a, 0)$, $(5, 0)$. Given that the system is in equilibrium, find F and a.

5   A horizontal uniform rod AB of length 60 cm and mass 3 kg is supported by two vertical strings attached at 5 cm from A and 15 cm from B. How far from A must a 5 kg particle be placed (a) to equalise the tensions in the strings, (b) just to cause one string to become slack?

6   Forces of $-5\mathbf{i}$, $F\mathbf{i}$, $-8\mathbf{i}$, $7\mathbf{i}$ act at the points $(0, -9)$, $(0, -3)$, $(0, 2)$, $(0, a)$. Find F and a, given that the system is equivalent to a clockwise couple of moment 32.

7 Forces of $F\mathbf{i} - 3\mathbf{j}$ and $-10\mathbf{i} + 7\mathbf{j}$ act at the points $(6, -1)$ and $(a, 3)$, respectively. Given that the anticlockwise moment about the origin is 6, and the moment about the point $(0, 3)$ is zero, find F and a.

8 A non-uniform rod AB of length $8a$ and weight $10W$ rests horizontally on supports distant $2a$ from A and a from B. When an anticlockwise couple of moment $15Wa$ acts on the rod, the forces on the supports are in the ratio $3 : 2$. Find (a) the distance of the centre of gravity from A, (b) the forces on the supports when an additional anticlockwise couple of moment $5Wa$ is applied.

9 Find the resultant of (a) a force of $5\mathbf{i}$ at the origin together with an anticlockwise couple of moment 35, (b) a force of $8\mathbf{j}$ at $(5, 0)$ together with a clockwise couple of moment 112.

10 Forces of $3\mathbf{i}$, $6\mathbf{i}$, $-4\mathbf{i}$ act at the points $(0, -4)$, $(0, 5)$, $(0, 9)$, together with a clockwise couple of moment 8. Find the resultant.

11 AB is a non-uniform rod of length 40 cm whose centre of gravity is 14 cm from A. The rod rests on supports distant 4 cm from A and 12 cm from B, and a 3 kg particle is placed at A. Given that the forces on the supports are in the ratio $7 : 2$, find the weight of the rod.

12 Uniform rods AB, BC, of lengths 60 cm, 75 cm and masses 3 kg and 4 kg, are smoothly jointed at B and rest horizontally on two supports, one of which is 15 cm from A. Find the distance from B of the second support, and the forces on the supports.

13 Forces of $P\mathbf{j}$ and $Q\mathbf{j}$ act at the points $(3, 0)$ and $(-2, 0)$, respectively. When a clockwise couple of 40 is added to the system the resultant acts at $(2, 0)$, and when an anticlockwise couple of 20 is added (to the original system) the resultant acts at $(-10, 0)$. Find P and Q.

14 Uniform rods AB, BC, of lengths 80 cm and 120 cm, are smoothly jointed at B and rest on supports distant 10 cm from A and 40 cm from B. Given that the total weight of the rods is 26 N, find their individual weights and the forces on the supports. If a couple of moment 140 N cm, tending to move B downwards, is applied to AB, what couple must be applied to BC to maintain equilibrium?

Exercise R7

1 Forces of 4 N, 7 N, 8 N, 2 N act along the sides AB, CB, CD, AD of a rectangle $ABCD$ in which $AB = 6$ cm and $BC = 4$ cm. Find the magnitude of the resultant, the tangent of the angle it makes with AB, and the point where its line of action meets AB.

2 $ABCD$ is a rectangle in which $AB = 12$ cm and $BC = 9$ cm. Forces of 17 N, 3 N, 10 N, 15 N act along BA, BC, DC, AC. Find the magnitude of the resultant, the tangent of the angle it makes with AB, and the point where its line of action meets AB.

3 ABC is a triangle in which $AC = 20$ cm, $\angle A = 90°$ and $\angle C = 70°$. D is a point on BC such that $\angle ADC = 80°$. Forces of 9 N, 3 N, 7 N, 16 N act along BA, AC, AD, BC. Find the point at which the line of action of the resultant meets (a) AB, (b) AC.

4 Forces of $3\mathbf{i} + 6\mathbf{j}$ and $5\mathbf{i} - 4\mathbf{j}$ act at the points $(4, 2)$ and $(-1, -2)$, respectively. Find the points at which the line of action of the resultant meets the x-axis and the y-axis.

5 $ABCD$ is a rectangle in which $AB = 10$ cm and $\angle ABD = 35°$. Forces of 9 N,

2 N, 6 N, 7 N act along AB, BC, CD, BD. Find the magnitude of the resultant and the perpendicular distance of its line of action from B.

6   Forces of $7\mathbf{i} - 2\mathbf{j}$ and $-3\mathbf{i} + 10\mathbf{j}$ act at the points $(5, 0)$ and $(-2, 6)$, respectively. Find the magnitude of the resultant and the equation of its line of action.

7 ABC is a triangle in which $\angle C = 90°$, $\angle A = 30°$ and $AC = 12$ cm. D is the foot of the perpendicular from C to AB. Forces of $6\sqrt{3}$ N, 12 N, 3 N, $9\sqrt{3}$ N act along AC, BC, BA, CD. Prove that the system is equivalent to a couple and find its moment.

8 ABC is a triangle in which $\angle A = 80°$ and $\angle B = 30°$. Forces of P and 12 N act along AC and AB. Given that the line of action of the resultant bisects BC, find P.

9 $ABCD$ is a rectangle in which $AB = 24$ cm and $BC = 7$ cm. Forces of 96 N, 42 N, 144 N, 28 N, 50 N act along AB, CB, CD, AD, AC. Prove that the system is in equilibrium.

10 $ABCD$ is a rectangle in which $\tan ABD = 0.75$. Forces of P, 9, 8, Q, S act along AB, BC, CD, AD, DB. Given that the system is in equilibrium, find P, Q and S.

11   Forces of 2, P, Q (P and Q both being positive) act along the sides BA, AC, BC of an equilateral triangle ABC. Given that the line of action of the resultant trisects AB at right angles, find P, Q and the magnitude of the resultant.

12   Forces of $8\mathbf{i} + 3\mathbf{j}$ and $-6\mathbf{i} - 5\mathbf{j}$ act at the points $(5, 2)$ and $(1, -3)$, together with an anticlockwise couple of moment 8. Find the magnitude of the resultant, the angle it makes with the y-axis, and the point at which its line of action meets the y-axis.

13 $ABCD$ is a rectangle in which $AB = 10$ cm and $\angle BAC = 30°$. Forces of 3 N, $5\sqrt{3}$ N, 16 N act along CB, CD, AC, together with a couple of moment 20 N cm in the direction $A \rightarrow B \rightarrow C$. Find the magnitude of the resultant and the point at which its line of action meets AB. What additional couple would make the line of action pass through B?

14   Forces are completely represented by $2\overrightarrow{AB}$ and $3\overrightarrow{BC}$, where A, B, C are the points $(1, 2)$, $(2, 5)$, $(4, 1)$. Express the resultant in terms of $\mathbf{i}$ and $\mathbf{j}$, and obtain the equation of its line of action.

15   Forces of $\mathbf{i} + P\mathbf{j}, Q\mathbf{i} - 2\mathbf{j}, 2\mathbf{i} + 4\mathbf{j}$ act at the points $(2, 9), (5, -7), (-3, 5)$. Given that the resultant acts along the line $y = 2x$, in the upward direction, find P and Q.

16 ABC is a triangle in which $\angle B = 90°$ and $\tan C = 0.75$. D is the foot of the perpendicular from B to AC. Forces of $P, 9, 12, Q$ act along AB, BC, CA, DB. Find P and Q if (a) the system is equivalent to a couple, (b) the resultant acts along AC.

17 $ABCD$ is a rectangle in which $AB = 8$ cm and $BC = 6$ cm. E is the foot of the perpendicular from D to AC. Forces of P, Q, 5 N, 10 N, 20 N act along BA, DA, CD, AC, DE, together with a couple of moment L in the direction $A \rightarrow B \rightarrow C$. Find P, Q and L, given that (a) the system is in equilibrium, (b) the resultant is a force of 50 N along DE.

18   Forces which are completely represented by $4\overrightarrow{AB}, 2\overrightarrow{CB}$ and $\overrightarrow{CA}$ act along the sides of a triangle ABC. Prove that the resultant is completely represented by $9\overrightarrow{DE}$, where D trisects AC and E trisects BA.

19 $ABCD$ is a quadrilateral. Forces act on a particle which are represented in magnitude and direction by $\overrightarrow{AB}, \overrightarrow{AD}, \overrightarrow{BC}, \overrightarrow{CD}$. Prove that the resultant is represented in magnitude and direction by $4\overrightarrow{EF}$, where E and F are the mid-points of AB and BD, respectively.

20 *ABCDEF* is a regular hexagon. Prove that the system of forces which is completely represented by $\overrightarrow{AB}, \overrightarrow{CD}, \overrightarrow{FE}, \overrightarrow{DA}$ is equivalent to a couple whose moment is one third of the area of the hexagon.

21 *ABC* is a triangle in which the medians *AP, BQ, CR* meet at *G*. Prove that the resultant of the system of forces represented by $4\overrightarrow{RC}, 2\overrightarrow{QB}, 3\overrightarrow{AB}$ is represented in magnitude and direction by $6\overrightarrow{AP}$.

Exercise R8

1 A 5 kg block moving at $12\,\mathrm{m\,s^{-1}}$ strikes a 3 kg block moving in the opposite direction at $4\,\mathrm{m\,s^{-1}}$, and the blocks then move together. Find their common speed and the loss of KE at the impact.

2 A 300 kg gun standing on smooth level ground fires a 150 g bullet at $60°$ to the horizontal and recoils along the ground at $20\,\mathrm{cm\,s^{-1}}$. Find the speed of the bullet and the impulse it receives.

3 A ball is dropped from a height of 180 cm onto horizontal ground. Given that *e* for the ball and the ground is $\frac{2}{3}$, find the height to which the ball rises.

4 A ball is rolled at $80\,\mathrm{cm\,s^{-1}}$ along smooth horizontal ground from a point *P* towards a vertical wall. It rebounds from the wall and returns to *P*, taking $3\frac{1}{2}$ s for the complete journey. Given that $e = \frac{3}{4}$, find the distance from *P* to the wall.

5 A 2 kg body moving upwards at $12\,\mathrm{m\,s^{-1}}$ collides and coalesces with a 1 kg body which is moving downwards at $9\,\mathrm{m\,s^{-1}}$. If the collision occurs 30 m above the ground, what further time does it take for the combined mass to reach the ground?

6 A 3 kg ball moving at $8\,\mathrm{m\,s^{-1}}$ collides directly with a 1 kg ball which is moving in the same direction at $2\,\mathrm{m\,s^{-1}}$. $e = \frac{1}{3}$. Find the speeds after the collision, the loss of KE and the impulse between the balls at the collision.

7 A 2 kg ball moving at $7\,\mathrm{m\,s^{-1}}$ collides directly with a 1 kg ball which is moving in the opposite direction at $5\,\mathrm{m\,s^{-1}}$. $e = \frac{1}{4}$. Find the speeds after the collision, the loss of KE and the impulse between the balls at the collision.

8 A 150 kg gun stands on a smooth plane which is inclined at arc $\tan \frac{3}{4}$ to the horizontal. It fires a bullet at $600\,\mathrm{m\,s^{-1}}$ horizontally, and recoils up the plane at $12\,\mathrm{cm\,s^{-1}}$. Find the mass of the bullet.

9 A 2 kg ball moving at $7\,\mathrm{m\,s^{-1}}$ collides directly with a 3 kg ball, and after the collision the 2 kg ball moves in the opposite direction at $6\frac{1}{2}\,\mathrm{m\,s^{-1}}$. Given that $e = \frac{1}{2}$, find the speeds of the 3 kg ball before and after the collision.

10 A 400 g hammer, moving horizontally, strikes a stationary 30 g nail, and rebounds from it at half its original speed. The nail then penetrates 0.4 cm into wood which provides an average frictional resistance of 600 kN. Find the original speed of the hammer.

11 A 5 kg ball moving at $12\,\mathrm{m\,s^{-1}}$ collides directly with a 3 kg ball which is moving in the opposite direction at $2\,\mathrm{m\,s^{-1}}$. After the collision the relative speed is $10\,\mathrm{m\,s^{-1}}$. Find the actual speeds after the collision and the value of *e*.

12 As a result of an internal explosion, a 12 kg body splits into two parts with masses of 8 kg and 4 kg. Given that the larger piece is brought to rest, and the KE supplied by the explosion is 4.8 kJ, find the original speed of the 12 kg body.

13 Three identical balls *A, B, C* lie at rest in a straight line on a smooth horizontal surface. $e = \frac{1}{3}$ for each pair of balls. When *A* is projected towards *B* with a speed of u, show that just three collisions occur and find the final speeds of the balls in terms of u.

14   A particle A of mass $2m$ directly strikes a particle B of mass m which is initially at rest on a smooth horizontal surface. B takes $5\,\text{s}$ to reach a point P, $70\,\text{cm}$ away, and after a further $9\,\text{s}$ A also reaches P. Calculate the initial speed of A and the value of e.

15   Two balls A and B, with masses in the ratio $1:2$, lie at rest on a smooth horizontal surface. A is projected towards B, and the latter then strikes a vertical wall and rebounds towards A. Prove that the condition for a second collision between the balls to occur is that $E > \dfrac{2e-1}{e+1}$, where e and E are the coefficients of restitution for the pair of balls and for B and the wall, respectively.

16   A ball of mass m, moving at $80\,\text{cm s}^{-1}$, collides directly with a ball of mass $4m$ which is initially at rest. The collision occurs on a smooth horizontal surface, midway between two vertical walls which are $1\,\text{m}$ apart. Given that collisions between the balls and the walls are perfectly elastic, and that $e = \frac{1}{2}$ for collisions between the balls, find the time interval between the first two collisions of the balls.

17   Two balls A and B, with masses in the ratio $2:3$, lie at rest on a smooth horizontal surface. A is projected towards B, and after their collision B strikes a vertical wall, rebounds and strikes A again, bringing A to rest. Given that e for the pair of balls is $\frac{1}{3}$, find e for B and the wall.

18   Balls A, B, C, with masses in the ratio $5:5:4$, lie at rest in a straight line on a smooth horizontal surface. e is the same for each pair of balls. Prove that, when A is projected towards B, the condition for more than two collisions to occur is that $2e^2 - 5e + 2 > 0$. Solve this inequality to find the range of possible values of e.

19   A ball is dropped onto a horizontal plane and strikes it at a speed of u. Prove that the ball takes a further time of $\dfrac{2ue}{g(1+e)}$ to come to permanent rest on the plane.

20   A ball is dropped from a height of h onto a horizontal plane. Prove that the total distance it travels before finally coming to rest on the plane is $\dfrac{h(1+e^2)}{1-e^2}$.

Exercise R9

1   A uniform sphere is held in contact with a smooth vertical wall by a string, equal in length to the radius, which connects a point on the surface to a point on the wall. Prove that the tension in the string is twice the reaction of the wall.

2   A $12\,\text{kg}$ particle is just on the point of sliding on a plane inclined at arc $\tan\frac{3}{4}$ to the horizontal. Find μ. If the inclination is increased to arc $\tan\frac{4}{3}$, what horizontal force will just prevent the particle from sliding?

3   A $5\,\text{kg}$ particle lies on a plane inclined at $40°$ to the horizontal. Given that a force of $60\,\text{N}$, acting away from the plane at $20°$ to a line of greatest slope, is just sufficient to move the particle up the plane, calculate the value of μ.

4   A uniform rod AB is smoothly hinged to a fixed support at A, and rests at $50°$ to the downward vertical owing to a string connecting B to a point C which is vertically above A. Given that $\angle ACB = 25°$, and the tension in the string is $40\,\text{N}$, find the weight of the rod and the reaction of the hinge.

5   A body resting on a horizontal plane can just be made to slide along the plane by a force of 80 N acting at 65° to the upward vertical. Given that the angle of friction is 35°, find the body's mass.

6 A uniform rod is smoothly hinged to a fixed support at its lower end, and is held at an angle of arc tan $\frac{3}{4}$ to the horizontal by a string which is perpendicular to the rod and attached at a distance a from the lower end. If the reaction of the hinge is horizontal, prove that the length of the rod is $25a/8$.

7 A uniform ladder rests in limiting equilibrium at 60° to the horizontal with its foot on rough horizontal ground and its upper end against a smooth vertical, wall. Given that the reaction of the wall is 65 N, find the weight of the ladder and the value of μ for the ladder and the ground.

8 A 600 g rod is smoothly hinged to a fixed point at its lower end, and is held at 78° to the upward vertical by a horizontal string which is attached to a point $\frac{2}{3}$ of the way up the rod. A particle is suspended from the upper end of the rod. Given that the reaction of the hinge acts along the rod, find the mass of the particle, the magnitude of the reaction, and the tension in the string.

9 Owing to a force of 70 N at its lower end, a uniform rod rests at 40° to the horizontal with its upper end in contact with a smooth vertical wall. Find the weight of the rod and the reaction of the wall.

10 A uniform ladder of weight W rests at 50° to the horizontal with its foot on rough horizontal ground and its upper end against a smooth vertical wall. Given that a man of weight $2W$ can climb $\frac{3}{4}$ of the way up the ladder before it slips, find the value of μ for the ladder and the ground.

11 $ABCD$ is a uniform rectangular lamina of weight 20 N in which $AD = 2AB$ and E is the mid-point of AD. The lamina can turn in a vertical plane about a smooth hinge at A, and it rests with AD horizontal, B being above A, owing to a force of magnitude F. Find F and the magnitude and direction of the reaction of the hinge if the force of F acts (a) along DB, (b) along EC.

12 A uniform cylinder whose height is twice its diameter stands with one of its circular faces in contact with a horizontal plane. A gradually increasing force is applied to a point P on the upper circular edge. The force is directed at 70° to the upward vertical and is tending to tip the cylinder about the point vertically below P. Given that the cylinder is on the point of tipping and the point of sliding at the same moment, find the value of μ for the cylinder and the plane.

13 A uniform rod ACB in which $AC = \frac{1}{4}AB$ rests at an angle of θ to the vertical with its upper end B against a smooth vertical wall. At C the rod is supported by a rough peg. Given that μ for the rod and the peg is 2, find the two possible values of $\tan\theta$ when the rod is in limiting equilibrium.

14 A non-uniform rod AB of weight 4 N and length 25 cm is smoothly hinged to a fixed support at its upper end A. A particle of weight 2 N is attached at B, and the rod rests in equilibrium at 15° to the horizontal owing to a force of F, at right angles to the rod, which acts at a point 20 cm distant from A. Given that the reaction R of the hinge is horizontal, find R, F and the distance of the centre of gravity of the rod from A.

15 A uniform ladder of weight W is on the point of slipping when resting at arc tan $\frac{4}{3}$ to the horizontal with its upper end against a smooth vertical wall and its foot on rough horizontal ground. When a man of weight $\frac{8W}{3}$ stands $\frac{3}{4}$ of the way up the ladder, find the extra horizontal force which is needed at the foot of the ladder to maintain equilibrium.

16 $ABCD$ is a light rod in which $AB:BC:CD = 2:3:1$. Particles of weight $2W$ and W are attached to the rod at B and D, respectively, and it rests in equilibrium at an angle of θ to the horizontal with its upper end A in contact with a smooth vertical wall. At C the rod is supported by a rough peg. Obtain expressions in terms of W and θ for the normal and frictional components of the force exerted by the peg, and hence obtain an expression for the minimum possible value of μ.

17   Two uniform rods AB and BC, with weights of $3W$ and W, respectively, are smoothly jointed at B and rest in a vertical plane with A and C in contact with rough horizontal ground. The lengths of the rods are such that $\angle BAC = 60°$ and $\angle BCA = 30°$. Find the magnitude and direction of the reaction at B and the minimum possible value of μ for the rods and the ground.

18 ACB is a uniform rod of weight $5W$ in which $AC = \frac{1}{4}AB$. The rod rests at arc $\tan \frac{4}{3}$ to the horizontal with its foot B on rough horizontal ground. At C the rod is supported by another rod, of weight $2W$, which is held in a vertical position by a horizontal force at its mid-point. Given that the coefficient of friction for the pair of rods is $\frac{1}{2}$, find the minimum possible values of the coefficient for each of the rods and the ground.

Exercise R10

Assume that all pulleys and strings are light and smooth, and all strings are inextensible.

1   Bodies of mass 10 kg and 40 kg are connected by a string passing over a pulley. Find the acceleration of the system, the tension in the string and the force on the pulley.

2   Bodies A and B, with masses of 2 kg and 6 kg, respectively, are connected by a straight string and lie at rest on a rough horizontal surface. $\mu = \frac{1}{4}$ for each body and the surface. A force of 60 N is applied to A in the direction BA. Find the acceleration and the tension in the string.

3 A 900 g body on a smooth horizontal table is connected by a string passing over a pulley at the edge of the table to a 600 g body which hangs vertically. Find the acceleration, the tension in the string and the force on the pulley.

4   Bodies A and B, with masses of 4 kg and 6 kg, respectively, are connected by a straight string and lie with B above A on a line of greatest slope of a smooth plane inclined at $30°$ to the horizontal. The system is released, and as it slides down the plane the frictional forces on A and B are 5 N and 10 N, respectively. Find the acceleration and the tension in the string.

5   A 6 kg body on a smooth horizontal table is connected to strings which pass over pulleys at opposite edges of the table. One string supports a 9 kg body, and the other supports a 1 kg pan which holds a 4 kg body. Find the tension in each string and the reaction between the pan and the 4 kg body.

6   Bodies A and B, with masses of 10 kg and 20 kg, respectively, are connected by a string and pulled along a rough horizontal surface by a force of F applied to B in the direction AB. The coefficient of friction is 0.4 for A and 0.6 for B, and the tension in the string is 70 N. Find the value of F.

7   A 500 g body on a smooth horizontal table is connected by a string passing over a pulley at the edge of the table to a body of mass m grams which hangs vertically. Find m if (a) the acceleration is $2\,\mathrm{m\,s^{-2}}$, (b) the tension in the string is 3 N.

8   The system of question 1 is released from rest, and after travelling 750 cm the

40 kg body is brought to rest. Find the further distance travelled by the 10 kg body before it comes to instantaneous rest.

9 A 9.5 kg body is 112 cm from the edge of a smooth horizontal table. It is connected by a string passing over the edge of the table to a 500 g body which hangs vertically. The system is released from rest, and after falling 64 cm the 500 g body comes to rest on the ground. Find the total time the 9.5 kg body takes to reach the edge of the table.

10 The system of question 1 is released from rest, and after 2 s the 10 kg body picks up a stationary 50 kg body. Find the further time that elapses before the system comes to instantaneous rest.

11 Bodies of mass 2 kg and 3 kg are connected by a string passing over a pulley. What mass must be added to the smaller body to double the force on the pulley?

12 A 15 kg body lies on a smooth plane inclined at arc $\tan \frac{3}{4}$ to the horizontal. It is connected by a string which passes over a pulley at the top of the plane to a body which hangs vertically. Find the mass of this body, given that (a) the 15 kg body remains stationary, (b) the 15 kg body accelerates up the plane at $4 \, \mathrm{m \, s^{-2}}$, (c) the tension in the string is 60 N.

13 Two pans, each of mass 1 kg, are connected by a string passing over a pulley and initially hang at rest in equilibrium. A 3 kg body is dropped onto one of the pans, striking it at $5 \, \mathrm{m \, s^{-1}}$, and $\frac{1}{2}$ s later a 5 kg body is dropped onto the other pan at $6 \, \mathrm{m \, s^{-1}}$. Find the further time that elapses before the pans return to their original positions.

14 A rough plane and a smooth plane, with inclinations to the horizontal of arc $\tan \frac{3}{4}$ and arc $\tan \frac{4}{3}$, respectively, meet along their top edges. Bodies of mass 30 kg and 50 kg stand on the respective planes and are joined by a string passing over a pulley at the top edge. (a) If the acceleration of the system is $2 \, \mathrm{m \, s^{-2}}$, find the coefficient of friction for the 30 kg body and the rough plane. (b) Find the mass which must be added to the 30 kg body to produce the same acceleration in the opposite direction.

Exercise R11

Assume that all strings are light, and all are inextensible unless otherwise stated.

1 Find the angular velocity in $\mathrm{rad \, s^{-1}}$ of a particle which (a) performs 200 revolutions every 5 minutes, (b) moves in a circle of radius 80 cm at $2 \, \mathrm{m \, s^{-1}}$.

2 A particle moving in a circle of radius 75 cm has a centripetal acceleration of $48 \, \mathrm{m \, s^{-2}}$. Find its actual speed and its angular speed.

3 A 2 kg particle moves in a circle at $3 \, \mathrm{m \, s^{-1}}$ under the influence of a centripetal force of 16 N. Find the radius of the circle.

4 A particle is placed on a turntable, 15 cm from the centre. If the coefficient of friction is $\frac{1}{4}$, what is the maximum angular speed at which the turntable can rotate without the particle slipping outwards?

5 A 5 kg particle is attached to one end of an elastic string of natural length 40 cm. The other end is attached to a point on a smooth table, and the particle describes horizontal circles of radius 50 cm about this point, taking 1.2 s for each. Find (a) the modulus of elasticity of the string, (b) the speed at which the particle would describe circles of radius 60 cm.

6   An elastic string of natural length *l* is stretched to a length of 2*l* by a particle hanging vertically. Prove that the speed at which the particle causes the same extension by describing horizontal circles on a smooth table is $\sqrt{2gl}$.

7   A conical pendulum consists of a string of length 120 cm which carries a 6 kg particle. Find the tension in the string and its inclination to the vertical when the pendulum rotates at 4 rad s^{-1}.

8   The string of a conical pendulum is 80 cm long and it rotates at an angle of 35° to the vertical. Find the speed of the particle.

9   A particle describes horizontal circles of radius 9 cm around the inside of a smooth hemispherical bowl of radius 15 cm which is fixed with its rim horizontal. Find the angular speed of the particle.

10   A particle describes horizontal circles around the inside of a smooth hemispherical bowl which is fixed with its rim horizontal. The reaction of the bowl is $1\frac{1}{2}$ times the particle's weight, and the speed of the particle is 2 m s^{-1}. Find the radius of the bowl.

11   A conical pendulum consists of a $1\frac{1}{2}$ kg particle on the end of an elastic string of natural length 80 cm and modulus 160 N. Find the speed of the particle when the extension of the string is 10 cm.

12   A string of length 140 cm is attached at its ends to two points 1 m apart in a vertical line. The string passes through a smooth ring of mass 3 kg which rotates in a horizontal circle at a distance of 80 cm from the upper end of the string. Find the speed of the particle and the tension in the string.

13   A string of length 1 m is attached at its ends to two fixed points in a vertical line which are 70 cm apart, and a particle attached to the mid-point of the string rotates in a horizontal circle. Find the angular speed of the particle when (a) the tensions in the two sections of string are in the ratio 2 : 1, (b) the tension in the upper section is 60 N and the mass of the particle is $3\frac{1}{2}$ kg.

14   The wheels of a car are 1.4 m apart and the height of the centre of gravity is 80 cm. The car travels on a level road of radius 60 m. Find the speed at which it overturns and, given that this occurs, the minimum possible value of μ.

15   A banked railway track is part of a circle of radius 400 m. The rails are 1.5 m apart and the outer rail is 24 cm above the inner rail. Find the speed at which trains should run if no lateral force is to be provided by the rails.

16   A circular road of radius 125 m is banked at arc tan $\frac{3}{4}$ to the horizontal. Find the maximum and minimum speeds at which a car can travel on the road without sliding, given that μ for the car and the road is $\frac{1}{2}$.

17   A car travels on a banked track of radius 300 m. Given that no lateral frictional force is required at a speed of 20 m s^{-1}, and that the car is on the point of sliding up the track at a speed of 40 m s^{-1}, find the value of μ.

Exercise R12

All the given velocities may be assumed to be constant. When distances or velocities are given in terms of **i** and **j**, the units are metres and m s^{-1}, respectively.

1   Two cars move at 90 km h^{-1} and 48 km h^{-1} in the directions due north and due west, respectively. Find the velocity of the first car relative to the second.

2   The velocities of particles *A* and *B* are 11**i** − 8**j** and 3**i** − 2**j**, respectively. Find (a) the velocity of *A* relative to *B*, (b) the speed of *A* relative to *B*.

3 To a man walking due south at $4\,\text{km}\,\text{h}^{-1}$ the wind appears to be coming from $80°$ east of south at the same speed. Find the true velocity of the wind.

4 Two particles start together and move with velocities of $12\mathbf{i}+15\mathbf{j}$ and $5\mathbf{i}-9\mathbf{j}$. Find their relative speed and their distance apart after $4\,\text{s}$.

5 The velocity of a particle P is $25\mathbf{i}-30\mathbf{j}$ and that of P relative to a particle Q is $35\mathbf{i}-45\mathbf{j}$. Find the velocity of Q.

6 A girl can swim at $75\,\text{cm}\,\text{s}^{-1}$ in still water. How long does she take to cross a river $90\,\text{m}$ wide which is flowing at $60\,\text{cm}\,\text{s}^{-1}$ if (a) the distance she travels is a minimum, (b) the time she takes is a minimum?

7 A particle P starts at the point $(-2,-2)$ and moves with a velocity of $8\mathbf{i}+9\mathbf{j}$, while a particle Q starts at the point $(4,0)$. Find the minimum distance between the particles if the velocity of Q is (a) $5\mathbf{i}+9\mathbf{j}$, (b) $8\mathbf{i}+6\mathbf{j}$, (c) $6\mathbf{i}+7\mathbf{j}$.

8 A particle A starts at the point $(0,2)$ and moves with a velocity of $11\mathbf{i}+k\mathbf{j}$, while a particle B starts at the point $(3,3)$ and moves with a velocity of $2\mathbf{i}-k\mathbf{j}$. Find the value of k such that the particles collide.

9 A particle P starts on the x-axis and moves with a velocity of $12\mathbf{i}-3\mathbf{j}$, while a particle Q starts at the point $(4,3)$ and moves with a velocity of $8\mathbf{i}-5\mathbf{j}$. Find the initial position of P if the particles collide.

10 One car drives due north at $30\,\text{m}\,\text{s}^{-1}$ while another car, initially $600\,\text{m}$ due north of the first, drives at the same speed on a bearing of $100°$. Find the time it takes (a) for the cars to reach the positions at which they are closest together, (b) for the second car to be due east of the first.

11 An aeroplane has a speed of $200\,\text{km}\,\text{h}^{-1}$ in still air. How long does it take to fly $500\,\text{km}$ due east if a $70\,\text{km}\,\text{h}^{-1}$ wind is blowing from north-east?

12 Two people are directly opposite each other on the banks of a river $200\,\text{m}$ wide which is flowing at $3\,\text{km}\,\text{h}^{-1}$. One person walks along the bank, against the flow of the water, at $6\,\text{km}\,\text{h}^{-1}$, while the other crosses the river in a boat whose speed in still water is $12\,\text{km}\,\text{h}^{-1}$. Find (a) the time the crossing takes if the people reach the same point at the same time, (b) their distance apart after the crossing if the crossing takes the least possible time.

13 A walker sets off at $1\frac{1}{2}\,\text{m}\,\text{s}^{-1}$ in a north-easterly direction, and at the same time a cyclist, initially $200\,\text{m}$ south-east of her, sets off with such a speed and direction that they meet after $40\,\text{s}$. Find the distances covered by the walker and the cyclist when they meet.

14 A ship S is initially $150\,\text{m}$ due north of a ship T. S steams due south at $10\,\text{m}\,\text{s}^{-1}$ while T steams on a bearing of $240°$ at $16\,\text{m}\,\text{s}^{-1}$. Find their distance apart after (a) $15\,\text{s}$, (b) $20\,\text{s}$.

15 A current is flowing at $5\,\text{km}\,\text{h}^{-1}$ on a bearing of $310°$. Find the minimum speed (relative to the water) at which a man must swim in order to travel (a) due north, (b) south-west.

16 A ship with a speed in still water of $35\,\text{km}\,\text{h}^{-1}$ has to travel north-west in a current running due west. Given that the pilot can achieve this by setting a course of $80°$, find the other course which would achieve the same effect and the resultant speed of the ship in each case.

17 A particle P starts at the point with position vector $-6\mathbf{i}$ and moves with a velocity of $-3\mathbf{i}+5\mathbf{j}$, while a particle Q starts at the point with position vector $5\mathbf{i}+3\mathbf{j}$ and moves with a velocity of $-5\mathbf{i}+4\mathbf{j}$. Find (a) the time that elapses before the particles are in a line parallel with the y-axis, (b) the time interval for which they are less than $3\,\text{m}$ apart.

18 Two cars are moving with speeds of $40\,\text{m}\,\text{s}^{-1}$ and $15\,\text{m}\,\text{s}^{-1}$ on roads which meet at 60°. Both cars are approaching the junction, and at a certain moment both are 200 m from it. Find their minimum distance apart.

19 A man swims across a river 60 m wide, reaching a point 80 m downstream from his original position. Given that the minimum speed (relative to the water) at which he has to swim to achieve this is $0.9 \, \text{m s}^{-1}$, find (a) the speed of the river. If he makes the return journey in a boat, and takes 50 s, find (b) the speed of the boat relative to the water.

20 A particle Q starts at the origin and moves with a velocity of $6\mathbf{i} + 4\mathbf{j}$, while a particle P moves with a velocity of $2\mathbf{i} + 7\mathbf{j}$. After 2 s the particles are at their minimum distance apart, namely 5 m. Find the two possible starting positions of P.

Exercise R13

When velocities are given in terms of $\mathbf{i}$ and $\mathbf{j}$, the units are m s^{-1}.

1 A body is projected with a velocity of $12\mathbf{i} + 25\mathbf{j}$. Find its velocity, its speed, and the inclination of its path to the horizontal, after 2 s.

2 A ball is thrown horizontally, with a speed of $20 \, \text{m s}^{-1}$, from the top of a building 45 m high. Find the time it takes to reach the ground and the distance from the foot of the building at which it does so.

3 A body is projected with a speed of $45 \, \text{m s}^{-1}$ at an inclination of 50° to the horizontal. Find the time it takes to return to its original level, its greatest height and its horizontal range.

4 Find (a) the maximum horizontal range of a body projected at $90 \, \text{km h}^{-1}$, (b) the time of flight of the body when achieving this range.

5 The maximum horizontal distance a boy can throw a ball is 40 m. If he throws the ball at maximum speed, but only achieves a range of 30 m, what are the two possible angles of projection?

6 After $1\frac{1}{2}$ s of flight the position vector of a projectile with respect to its initial position is $42\mathbf{i} + 12.75\mathbf{j}$. Find the initial velocity of the projectile and its position vector after a further $\frac{1}{2}$ s.

7 Find (a) the angle of projection of a projectile whose horizontal range is 3 times its greatest height. With this angle of projection, find (b) the initial speed of the projectile if its time of flight is 5.6 s.

8 The initial velocity of a projectile is $4\mathbf{i} + 25\mathbf{j}$. Find the two times at which it is moving at arc $\tan 2\frac{1}{2}$ to the horizontal.

9 A body is projected with a speed of $30 \, \text{m s}^{-1}$ at an angle of arc $\tan 2$ to the horizontal. Find the height of the body above its starting point when it has travelled 45 m horizontally.

10 The initial speed and elevation of a projectile are $40 \, \text{m s}^{-1}$ and 30°, respectively. Find (a) the time it takes to reach a point 60 m below its starting point, (b) the horizontal distance it travels while more than 15 m above its starting point.

11 A body projected at 45° to the horizontal has a maximum height of H and takes a time of T to return to its original level. If the body is projected at the same speed, find the angles of projection which are such that (a) the greatest height is $4H/3$, (b) the horizontal range is $2H$, (c) the time the body takes to return to its original level is $T/\sqrt{2}$.

12 A ball thrown at $30 \, \text{m s}^{-1}$ has to pass through a point which is 5 m above the point of projection and a horizontal distance of 60 m from it. Find the two possible angles of projection.

13 A projectile with an initial speed of V has the same horizontal range with two different angles of projection. If H_1 and H_2 are the greatest heights of the two flights, prove that $H_1 + H_2 = V^2/2g$.

14 A ball is thrown from ground level, strikes a vertical wall at right angles at a height of 35 m, rebounds, and strikes the ground again at a distance of 30 m from the wall. Given that the coefficient of restitution for the ball and wall is $\frac{3}{4}$, find the ball's initial speed and inclination to the horizontal.

15 A body is projected from ground level with a speed of 55 m s^{-1}. It just clears a wall of height 40 m which is 220 m from the point of projection, then strikes the ground at a point beyond the wall. Find the two possible values of the distance from this point to the wall.

16 Balls are thrown at the same moment from the tops of two buildings which are 40 m apart, and they collide at a point on the ground between the buildings. The first building is 8 m high, and the ball thrown from it has an initial speed of 10 m s^{-1} at an angle of arc tan $\frac{4}{3}$ to the upward vertical. The second ball is thrown horizontally. Find the speed of the second ball and the height of the second building.

17 A body is projected from ground level with a speed of 36 km h^{-1} at an angle of arc tan 2 to the horizontal. It just clears two posts, each of which is 3 m high. Find the distances of the posts from the point of projection.

18 A ball is thrown with a speed of 30 m s^{-1} at $60°$ to the horizontal, the point of projection being a distance of $15\sqrt{3} \text{ m}$ from a smooth vertical wall. The ball strikes the wall, rebounds, and returns to its starting point. By considering the horizontal motion, find the time the ball takes to reach the wall, and by considering the vertical motion find the total time of flight. Deduce the value of the coefficient of restitution for the ball and the wall.

19 When a body is thrown at $45°$ to the horizontal the highest point in its flight is P. Find the other angle of projection such that the body passes through P when thrown with the same initial speed.

20 A body is projected with a speed of V at an angle of α to the horizontal. Prove that the direction of motion is at right angles to the original direction after a time of $\dfrac{V}{g \sin \alpha}$.

21 A particle is projected with an initial velocity of $16\mathbf{i} + 24\mathbf{j}$, and 2 s later another particle is projected from the same point. Given that the particles collide at 36 m below the point of projection, find the initial velocity of the second particle.

22 One particle is projected from a point P with a speed of $5u$ at an angle of arc tan $\frac{3}{4}$ to the upward vertical, while another particle is projected simultaneously from a point Q which is a horizontal distance of $5u^2/g$ from P and at a different height. Given that the particles collide when each is moving at $45°$ to the upward vertical, find the initial speed of the second particle and the difference in height of P and Q.

23 A body is projected with a speed of V at an elevation of α, and at the same moment another particle is released from rest. If the particles collide when their speeds are equal, prove that (a) the collision occurs after a time of $\dfrac{V}{2g \sin \alpha}$, (b) the second particle starts at a height of $\dfrac{V^2}{2g}$ above the first.

Exercise R14

1   Particles of mass 5 kg, 3 kg, 2 kg are placed at the points $(1, 4)$, $(-7, 6)$, $(3, -4)$, respectively. Find the position of their centre of gravity.

2   Particles of mass 12 kg, 8 kg are placed at the points $(3, 6)$, $(-2, -4)$, respectively, and a particle of mass 5 kg is placed at the point (h, k). The centre of gravity is at the point $(4, -2)$. Find h and k.

3   An isosceles triangle CED is attached to a rectangle $ABCD$ to form a uniform lamina $ABCED$. $AB = 32$ cm, $BC = 18$ cm and $CE = 20$ cm. Find the distance of the centre of gravity from AB.

4 $ABCD$ is a uniform rectangular lamina in which $AB = 18$ cm, $BC = 15$ cm and E is the mid-point of AB. The triangle BCE is removed. Find (a) the distances of the centre of gravity of the remaining lamina from AD and DC, (b) the inclination of AD to the vertical when this lamina is suspended from C.

5 $ABCD$ is a uniform square lamina of side 10 cm. E, F are points on AB, BC, respectively such that $BE = BF = 4$ cm. The square $BEGF$ is removed and superimposed on the lamina so that two of its sides are coincident with AD and DC. Find the distance of the centre of gravity from D.

6 $ABCD$ is a piece of uniform wire of weight 5 N which is smoothly hinged to a fixed support at A. $\angle ABC = \angle BCD = 90°$, $AB = BC = 40$ cm and $CD = 20$ cm. Find the minimum force at D such that the wire hangs with (a) AB vertical, (b) AB horizontal.

7 $ABCD$ is a uniform lamina in the shape of a trapezium in which AB is parallel to DC, $\angle A = \angle D = 90°$, $AB = AD = 8$ cm and $DC = 2$ cm. The mass of the lamina is 5 kg, and particles of mass 4 kg and 1 kg are attached to it at A and B, respectively. When the lamina is placed in a vertical plane with DC on horizontal ground, find (a) the vertical force at B which is required to prevent toppling about C, (b) the minimum force at B which has this effect.

8 G is the centre of gravity of an L-shaped piece of uniform wire ABC in which $\angle B = 90°$. Prove that $\tan ABG = (\tan BAC)^2$.

9   A uniform lamina in the form of an equilateral triangle ABC is held in a vertical plane with A vertically above B by a horizontal string attached at A and a string attached at C. Prove that the ratio of the tensions in the strings is $2 : \sqrt{7}$.

10   The lamina of question 3 is suspended from A. Given that the mass of the lamina is 8 g, find the mass of the particle which must be attached at B to make the lamina hang with AC vertical.

11   A semicircular piece of uniform wire has its two ends joined by a straight piece of uniform wire of twice the density. This body is suspended from one of the ends of the straight wire. Find the angle between the straight wire and the vertical.

12   Find the centroids of the regions bounded by (a) the graph of $y = 3x - x^2$ and the x-axis, (b) the graph of $y = \sin x$, the x-axis, the origin and the line $x = \dfrac{\pi}{2}$.

13   From a uniform solid cylinder of radius R and height $3R$ a solid right cone of radius R and height $2R$ is removed, the plane face of the cone being one of the original plane ends of the cylinder. The remaining body is placed with its circular face on a horizontal plane, and this plane is steadily tilted. Assuming that the body does not slide, find the inclination of the plane to the horizontal when it is on the point of toppling.

14   A uniform solid right cone of semi-vertical angle arc $\tan \frac{1}{2}$ is suspended by two vertical strings, one attached to a vertex and one to a point on its circular face. If it hangs with its upper slant edge horizontal, what is the ratio of the tensions in the strings?

15 The region bounded by the graph of $y^2 = 2x$ and the line $x = 2$ is completely rotated about the x-axis. Show that the x co-ordinate of the centroid of the solid of revolution formed is $1\frac{1}{3}$. A uniform solid of this form is attached to a uniform right cone of equal density, the two circular faces being coincident. Prove that if the centre of gravity of the compound body lies on the common circular face, the height of the cone is $\sqrt{8}$.

16 A compound body is formed by joining the circular faces of a uniform solid cylinder and a uniform solid right cone with the same radius. The height of the cylinder is k times that of the cone, and the density of the cone is k times that of the cylinder. When the body is placed on horizontal ground, show that it rests with the cylinder or the cone in contact with the ground according to whether k is greater than or less than $\frac{1}{6}$.

17 From a uniform solid hemisphere of radius r, a solid hemisphere of radius kr $(k < 1)$ is removed, the plane faces of the hemispheres and the centres of these faces being coincident. Prove that the distance of the centre of gravity of the remaining solid from the centre of the plane faces is

$$\frac{3r(1+k)(1+k^2)}{8(1+k+k^2)}.$$

Show that it follows that the centre of gravity of a hollow hemisphere of negligible thickness is a distance $r/2$ from the centre.

18 A hollow cylinder, closed at its lower end, is made of thin uniform metal. At its upper end it is attached to a hollow hemisphere of equal radius made of the same metal, thus forming a completely enclosed container with a circular base and a hemispherical top. Prove (a) that the height of the centre of gravity of the container is $\dfrac{(h+a)^2}{3a+2h}$, (b) that the centre of gravity is level with the lower edge of the hemisphere if $h = \dfrac{a}{2}(\sqrt{5}-1)$.

Exercise R15

Assume that coins and dice are 'fair' unless otherwise stated.

1 A card is drawn at random from an ordinary pack and a die is thrown. Find the probability of (a) a heart and a 3, (b) a red ace and an odd number, (c) neither a club nor a number over 4.

2 A number is chosen at random from each of the sets $\{2, 5, 6, 8\}$ and $\{6, 7, 8, 9, 10\}$. Find the probability that (a) the same numbers are chosen, (b) the sum of the numbers is 15, (c) the sum of the numbers is odd.

3 A and B are independent events. Given that $p(B') = \frac{1}{3}$ and $p(A \cap B) = \frac{5}{9}$, find $p(A)$ and $p(A \cup B)$.

4 A biased die is thrown 600 times, and a 6 occurs 240 times. Estimate the probability that at least one 6 is obtained in 3 throws.

5 In a class of 30 girls there are 25 who like at least one of the sports tennis and hockey. There are 14 who dislike hockey, and the number who like tennis is 5 more than the number who like hockey. Draw a Venn diagram and find the probability that (a) a girl chosen at random likes exactly one of the sports, (b) two girls chosen at random both dislike tennis.

6 Four equally matched teams A, B, C, D reach the semifinal of a knockout competition, the draw for the semifinal not yet having been made. Find the probability that (a) A beats B in the final, (b) both B and C fail to reach the final, (b) A plays one more game and D plays two.

7 A greyhound's chance of winning a 6-dog race is $\frac{1}{2}$ if he is drawn in trap 1, $\frac{1}{3}$ if he is drawn in traps 2, 3 or 4, and $\frac{5}{8}$ if he is drawn in traps 5 or 6. If the trap is chosen at random, what is the greyhound's overall chance of winning the race?

8 Three married couples are seated at random in a straight line. Find the probability that (a) the men sit together and the women sit together, (b) each man sits next to his wife.

9 Every day I go to work either by bus or by car. After going by bus the probability that I change my mode of transport on the following day is $\frac{2}{3}$, while if I go by car the probability that I change is $\frac{1}{2}$. Given that I go by bus on Monday, draw a tree diagram for the days up to and including Thursday, and find the probability that (a) I go by bus on Thursday, (b) my mode of transport is the same on Tuesday and Wednesday, (c) during the 4-day period my mode of transport is the same for 3 or more successive days.

10 When Tom plays Harry at table-tennis, the probability that Tom wins any particular point is $\frac{3}{5}$ and the probability that Harry wins the point is $\frac{2}{5}$. If the score reaches 20-all, what is the probability that (a) the game ends 22-20, (b) the score reaches 22-all, (c) Harry wins 23-21? (The game ends if either player achieves a 2-point lead.)

11 Events A, B are independent and events A, C are mutually exclusive. Given that $p(A) = \frac{1}{3}$, $p(A \cap B) = \frac{1}{5}$, $p(A \cup C) = \frac{1}{2}$ and $p(B \cap C) = \frac{1}{10}$, find $p(B)$ and $p(C)$ and prove that B and C are independent.

12 Three cards are chosen at random from a reduced pack consisting of the aces, kings, queens and jacks of each suit. Find the probability that (a) all the cards are of the same value (e.g. all kings), (b) exactly two of them are of the same value.

13 A coin is thrown 7 times. Find the probability that the number of heads is (a) 2, (b) 4.

14 A number is chosen at random from each of the sets $\{1, 2, 3\}$ and $\{5, 6, 7, 8\}$. In each of the following cases determine whether the given two events are independent.
(a) The total is 9. The first number is 3.
(b) The total is odd. The second number is 7.
(c) The total is odd. The first number is 2.
(d) The difference between the numbers is 4. The second number is 6.
(e) The total is more than 9. The first number is 2.

15 A bag contains three cards labelled 1 and two cards labelled 2. A pair of cards is drawn from the bag at random and the sum of the numbers on the cards is recorded. This is called the score. Now any cards labelled 2 are replaced in the bag and the procedure is repeated. Draw a tree diagram showing the probabilities of the various possible scores on the two draws, and find the probability that (a) the score on the second draw is 3, (b) the sum of the two scores is at least 7.

16 In a certain country 1 man in 12 plays cricket but not golf, 1 in 8 plays golf but not cricket, and 3 in 4 play neither cricket nor golf. Find the probability that a man chosen at random plays both sports, and determine whether playing cricket and playing golf are independent characteristics.

17   Four 4-sided dice, each with the sides numbered from 1 to 4, are thrown together. Find the probability that (a) all four numbers are obtained (in any order), (b) two 3's and two 4's are obtained.

18   Given that 1 person in 4 is left-handed, find, to 2 decimal places, the probability that a group of 6 people contains (a) exactly 1 left-handed person, (b) at least 2 left-handed people.

19   With the data of question 7, find the probability that the greyhound is (a) drawn in traps 5 or 6, given that he wins the race, (b) drawn in trap 1, given that he loses, (c) drawn in trap 3, given that he wins.

20   With the data of question 9, find the probability that (a) I travel by car on Tuesday, given that I travel by bus on Thursday, (b) I travel by bus on Wednesday, given that I travel by bus more often than by train over the 4-day period.

21   One box contains three cards labelled 1, 1, 2, and another box contains four cards labelled 1, 1, 2, 2. One card is drawn at random from each box and it is noted whether the sum of the numbers on the cards is odd or even. Without replacement of the cards drawn, this procedure is repeated. Find the probability that the two sums obtained are both odd.

22   A boy and a girl throw a coin in turn, starting with the girl. Show that the probability that the girl is the first to throw a head is the sum to infinity of the geometric series $\frac{1}{2} + \frac{1}{8} + \frac{1}{32} + \ldots$. Evaluate this probability, using the formula $S_\infty = \dfrac{a}{1-r}$.

23   A boy and a girl throw a die in turn, starting with the boy. Find the probability that the boy is the first to throw a 5 or a 6.

24   With the data of question 10, find the probability that Harry wins the game. (The winner is the first player to achieve a 2-point lead.)

Exercise R16

The units are SI throughout.

1   A particle starts with a velocity of $6\,\mathrm{m\,s^{-1}}$ and moves with an acceleration of $1 - 2t$. Find (a) its velocity after 2 s, (b) the time at which it is instantaneously at rest, (c) its maximum velocity.

2   A particle starts at the point $(2, 0)$ and moves along the x-axis with a velocity in the positive direction of $1/x$. Find (a) its position after 6 s, (b) the time it takes to reach the point $(6, 0)$.

3   A particle moves according to the equation $\mathbf{r} = (3t - 6)\mathbf{i} + t^2\mathbf{j}$. Find (a) the speed and direction when it crosses the y-axis, (b) the equation of the path.

4   A particle starts at rest and moves with an acceleration of $50\cos 5t$. Find the time at which it returns to rest for the first time and its displacement from the starting point at this time.

5   A 500 g particle moves according to the equation $\mathbf{r} = (t^4 - t^3)\mathbf{i} + (t^3 + t^2)\mathbf{j}$. Find the magnitude of the force on the particle (a) when $t = 1$, (b) at the two times when the force is parallel to the y-axis.

6   A particle starts at the origin with a velocity of $2\,\mathrm{m\,s^{-1}}$ and moves along the x-axis with an acceleration of $2x + 6$. Find (a) the velocity at the point $(3, 0)$, (b) the point at which the velocity is $6\,\mathrm{m\,s^{-1}}$.

7 A particle moving with an acceleration of $3t/2$ travels 20 m in the interval between $t = 2$ and $t = 4$. Find its velocity when $t = 2$.

8 A particle moves according to the equation $\mathbf{r} = (2t + \sin 2t)\mathbf{i} - (\cos 2t)\mathbf{j}$. Show that its speed is $|4 \cos t|$ and the inclination of its direction to the x-axis is t radians.

9 A particle starts at the point $(2, 0)$ with a velocity of $3\,\mathrm{m\,s^{-1}}$ and moves along the axis Ox with an acceleration of $6x^2$ towards O. Find (a) its speed at O, (b) its maximum distance from O.

10 A particle starts at the origin with a velocity of $5\mathbf{j}$ and moves with a constant acceleration of $4\mathbf{i}$. Find (a) the velocity and displacement after 2 s, (b) the speed after 3 s, (c) the time at which the particle crosses the line $x + y = 3$.

11 A particle starts with a velocity of $2\,\mathrm{m\,s^{-i}}$ and an acceleration of $0.75\,\mathrm{m\,s^{-2}}$, and moves such that its acceleration is inversely proportional to $v + 2$. Find (a) the time it takes for the velocity to increase from $3\,\mathrm{m\,s^{-1}}$ to $5\,\mathrm{m\,s^{-1}}$, (b) the velocity after 8 s.

12 The velocities of two observers A and B are $3t\mathbf{i} - (t^2 + 12)\mathbf{j}$ and $(t + 8)\mathbf{i} - 7t\mathbf{j}$, respectively. Write down an expression for the velocity of A relative to B and find the times at which A appears to B to be (a) instantaneously stationary, (b) moving parallel to the x-axis.

13 A 2 kg particle starts at the point $(2, 0)$ with a velocity of $-4\mathbf{j}$ and moves under the influence of a force of $6\mathbf{i} + 4t\mathbf{j}$. Find its position when it is moving (a) parallel to the x-axis, (b) at $45°$ to the positive x-axis.

14 A particle starts with a velocity of $6\,\mathrm{m\,s^{-1}}$ and moves with a velocity which is proportional to e^{-2t}. Show (a) that it approaches a limiting position which is 3 m from its starting point, (b) that it is 1 m from this limiting position after $\frac{1}{2}\ln 3$ seconds.

15 A 500 g particle starts at the point $(-8, -12)$ with a velocity of $6\mathbf{j}$ and moves under the influence of a force of $3t\mathbf{i}$. Show that the particle passes through the origin. If the force is removed when the particle reaches the origin, what will be its position after a further 3 s?

16 One particle starts at the point $(-3, 0)$ and moves such that $\mathbf{v} = 8t\mathbf{i} + 2\mathbf{j}$, while another starts at the point $(7, 1)$ and moves such that $\mathbf{v} = -\mathbf{i} + 2\mathbf{j}$. Obtain the equations of their paths. Find the point of intersection of the two paths, and hence show that the particles collide if the second starts $\frac{1}{2}$ s after the first.

17 A particle starts at the origin with a velocity of $2\,\mathrm{m\,s^{-1}}$ and an acceleration of $8\,\mathrm{m\,s^{-2}}$, and moves along the x-axis with an acceleration which is proportional to the square of its velocity. Prove that $v = 2e^{2x}$ and that $x = \frac{1}{2}\ln\dfrac{1}{1-4t}$. Hence or otherwise find the time at which the velocity is $4\,\mathrm{m\,s^{-1}}$.

18 A 3 kg particle which is initially moving at $30\,\mathrm{m\,s^{-1}}$ experiences a constant retarding force of 6 N until the speed is $12\,\mathrm{m\,s^{-1}}$, after which the retarding force becomes inversely proportional to the speed. Find the total time it takes for the particle to come to rest and the total distance it travels in that time.

19 A body of mass m is projected vertically upwards with a speed of V_o, and in addition to its own weight experiences a retarding force of mv. Prove that its greatest height is $V_o - g\ln\dfrac{V_o + g}{g}$ and that it reaches this height after a time of $\ln\dfrac{V_o + g}{g}$.

20 The acceleration of a particle is $e^{v/2}$. Prove that the distance it travels in accelerating from $2\,\mathrm{m\,s}^{-1}$ to $4\,\mathrm{m\,s}^{-1}$ is $\dfrac{4}{e^2}\,(2e-3)$.

21 Two particles move according to the equations

$$\mathbf{r} = (3\cos t)\mathbf{i} + 2\mathbf{j}, \qquad \mathbf{r} = (2\cos t)\mathbf{i} + (\sin t)\mathbf{j}.$$

Find the times at which the particles are first moving (a) directly towards each other, (b) directly away from each other. Show also that each particle moves relative to the other in a circle at constant speed, and find (c) their maximum and minimum distances apart.

22 A particle is released from rest in a viscous liquid which provides a resistance which is proportional to the square of the speed. Prove that if the terminal velocity is V_T, the distance travelled by the particle in accelerating from a speed of $V_T/4$ to a speed of $3V_T/4$ is $\dfrac{V_T{}^2}{2g}\ln\dfrac{15}{7}$.

Exercise R17

1 A particle describes an SHM in which the period and amplitude are 4 s and 80 cm, respectively. Find the maximum speed, the maximum acceleration and the speed of the particle when 30 cm from the central point.

2 A particle moving with SHM performs 20 complete oscillations, of amplitude 1.4 m, each minute. Find the distance of the particle from the central point, and its speed, 2 s after it leaves the central point.

3 A particle describing SHM has a speed of $3\,\mathrm{m\,s}^{-1}$ when its acceleration is zero, and an acceleration of $6\,\mathrm{m\,s}^{-2}$ when its speed is zero. Find (a) the amplitude and period of the motion, (b) the average speed, (c) the speed of the particle $\frac{1}{2}$ s after it leaves a position of instantaneous rest.

4 A particle moving with SHM has speeds of $4\,\mathrm{m\,s}^{-1}$ and $3\,\mathrm{m\,s}^{-1}$ when its distances from the central point are 50 cm and 1 m, respectively. Find the amplitude and period of the motion.

5 A particle starts at rest and moves with SHM. After 3 s it has a speed of $4\,\mathrm{m\,s}^{-1}$, and after another 2 s it is at rest again for the first time. Find its speed after a further 1 s.

6 At a time of 0.2 s after leaving its central point, a particle moving with SHM is 25 cm from that point and has an acceleration of $80\,\mathrm{cm\,s}^{-2}$. Find (a) the amplitude, (b) the speed of the particle after it has travelled a further 15 cm.

7 A particle oscillates with SHM between two points A and B. The central point is O, and P, Q are the mid-points of AO, BO, respectively. Find in terms of the period T the times taken for the journeys POQ and PAO.

8 A particle moving with SHM has a speed of V at points P and Q, and a speed of $V\sqrt{2}$ at the point which divides PQ in the ratio $1:3$. Prove that the amplitude of the motion is $\dfrac{PQ\sqrt{7}}{4}$ and the maximum speed is $V\sqrt{\dfrac{7}{3}}$.

9 A 3 kg particle is attached to one end of an elastic spring of natural length 2 m and modulus 50 N, and the other end of the spring is attached to a point P on a smooth horizontal table. The particle is released at a point on the table which is

3 m from P. Find (a) the period of the motion, (b) the speed of the particle 0.3 s after being released.

10   A simple pendulum is of length 70 cm and its maximum inclination to the vertical is $12°$. Find (a) the maximum speed of the bob, (b) its speed when the string is inclined at $4°$ to the vertical.

11   An elastic string of natural length 60 cm is extended to 70 cm by a force of 25 N. When a 3 kg particle is suspended by the string and set in vertical oscillation, find to the nearest whole number how many complete oscillations occur in each minute.

12   A body attached to the end of a vertical elastic string makes 45 complete oscillations, of amplitude 6 cm, every 30 seconds. Find the extension of the string when (a) the body is in the equilibrium position, (b) it is above the equilibrium position and moving at $50\,\mathrm{cm\,s^{-1}}$.

13   A simple pendulum of length 50 cm is held with the string at $5°$ to the vertical and the bob is projected towards its lowest point, at right angles to the string, with a speed of $40\,\mathrm{cm\,s^{-1}}$. Find (a) the amplitude of the subsequent oscillations, (b) the time the bob takes to come to rest for the first time.

14   A vertical spring carries a pan of mass m which extends it a distance of $2d$ when in equilibrium. The pan is pulled down a further distance of $5d$, then a body of mass $4m$ is placed on the pan and the system is released from rest. Prove that the speed of the pan when it has fallen a further distance of d is $\sqrt{\dfrac{gd}{2}}$.

15   An elastic string of natural length 80 cm and modulus 30 N is attached at one end to a fixed point P and at the other to a 250 g particle. The particle is held at P and released. Find the time it takes to reach its lowest point for the first time.

16   A particle suspended by a vertical spring extends the spring a distance of $2e$ when in equilibrium. The particle is pulled down a further distance of e, and released. When it passes through its equilibrium position, it picks up another particle, previously at rest, of half its mass. Prove that the amplitude of the oscillations which then occur is $e\sqrt{\dfrac{5}{3}}$.

Answers to Exercises

Exercise 1a (p. 6)

1 $22\,\mathrm{m\,s^{-1}}$, 48 m
2 $5\,\mathrm{m\,s^{-2}}$, 40 m
3 $5\,\mathrm{m\,s^{-2}}$, 15 m
4 175 m, $20\,\mathrm{m\,s^{-1}}$
5 2 s, $20\,\mathrm{m\,s^{-1}}$
6 $1\frac{1}{4}$ m, $5\,\mathrm{m\,s^{-1}}$
7 $5\,\mathrm{m\,s^{-2}}$, 4 s
8 36 m
9 (a) 1.6 s, (b) 4 m
10 3 s
11 $1.5\,\mathrm{m\,s^{-1}}$, $16.5\,\mathrm{m\,s^{-1}}$
12 20 m
13 $6\,\mathrm{m\,s^{-1}}$
14 0.6 s, 1 s
15 0.2 s
16 $10\,\mathrm{m\,s^{-1}}$, 2 s
17 10 m
18 $8\,\mathrm{m\,s^{-2}}$, 44 m
19 $1\frac{2}{3}\,\mathrm{m\,s^{-2}}$
20 1 s
21 1.8 m
22 2 s
23 $4\,\mathrm{m\,s^{-2}}$
24 27 m
25 5 s
26 $30\,\mathrm{m\,s^{-1}}$
27 $23\frac{1}{3}\,\mathrm{m\,s^{-1}}$, $1\frac{1}{3}\,\mathrm{m\,s^{-2}}$, $2\frac{1}{2}$ s
28 $0.8\,\mathrm{m\,s^{-2}}$

Exercise 1b (p. 9)

1 $2\frac{2}{3}\,\mathrm{m\,s^{-2}}$
2 $1\,\mathrm{m\,s^{-2}}$, $2\,\mathrm{m\,s^{-2}}$
3 3 s
4 12 m
5 $3\,\mathrm{m\,s^{-2}}$
6 $4\,\mathrm{m\,s^{-1}}$, $6\,\mathrm{m\,s^{-2}}$
7 $10\,\mathrm{m\,s^{-1}}$
8 5 s
9 15 m
10 2 s
11 $4\,\mathrm{m\,s^{-2}}$
12 165 m.

Exercise 2 (p. 17)

1 (a) 650 N, (b) 0.35 N
2 (a) 4.5 kg (b) 20 g
3 $2\,\mathrm{m\,s^{-2}}$, $6\,\mathrm{m\,s^{-1}}$
4 5 N 5 $10/9\,\mathrm{m\,s^{-2}}$, 1.8 kg 6 1.2 N
7 10 N 8 $12\,\mathrm{m\,s^{-1}}$
9 $3g/4$ 10 $3\frac{1}{3}\,\mathrm{m\,s^{-2}}$
11 $5\,\mathrm{m\,s^{-2}}$
12 $4g/3$
13 (a) 3.25 N (b) 1 N
14 $2.5\,\mathrm{m\,s^{-2}}$
15 $g/4$ 16 $x/y - g$
17 $1\frac{2}{3}$ kg 18 5 kg
19 10 N
20 4 N
21 (a) 150 N (b) 900 N (c) zero 22 900 N, 600 N
23 400 N 24 16 kN
25 $2W/g$ 26 $5\,\mathrm{m\,s^{-2}}$, $3\frac{3}{4}$ N 27 $133\frac{1}{3}$ N
28 (a) 4.5 N (b) 3 N
29 $8W$ 30 (a) 18 N (b) 15 N 31 (a) 35 N (b) 350 N
32 400 kN
33 2.6 N, 2 N, 1.2 N
34 22.5 N, 12 N
35 (a) 18 mg (b) 10 mg
36 75 N 37 160 N
38 64 N 39 20 cm²
40 $5\,\mathrm{m\,s^{-1}}$
41 0.8 N 42 50 cm²
43 $2\sqrt{15}\,\mathrm{m\,s^{-1}}$.

Exercise 3 (p. 23)

1 (a) $\overrightarrow{AD}$ (b) $\overrightarrow{EB}$ (c) $\overrightarrow{BC}$ (d) $\overrightarrow{DB}$ (e) $\overrightarrow{BC}$ (f) $\overrightarrow{BE}$
2 (a) $\mathbf{p} + \mathbf{s}$ (b) $-\mathbf{s}$ (c) $\mathbf{s} - \mathbf{p}$ (d) $-\frac{1}{2}(\mathbf{p} + \mathbf{s})$ (e) $\frac{1}{2}(\mathbf{p} - \mathbf{s})$
3 (a) $-2\mathbf{p}$ (b) $2(\mathbf{p} + \mathbf{s})$ (c) $\mathbf{p} - \mathbf{s}$ (d) $-2\mathbf{p} - \mathbf{s}$
4 (a) $\mathbf{i} + 4\mathbf{j}$ (b) $2\mathbf{i} - 3\mathbf{j}$ (c) $6\mathbf{i} + 5\mathbf{j}$ (d) $-3\mathbf{i} - 4\mathbf{j}$
5 (a) 5, 36.87° (b) $2\sqrt{2}$, 135° (c) 26, $-22.62°$ (d) $3\sqrt{13}$, $-123.7°$
6 (a) $-5\mathbf{i} + 4\mathbf{j}$ (b) $-2\mathbf{i} + 9\mathbf{j}$ (c) $17\mathbf{j}$ (d) $29\mathbf{i} - \mathbf{j}$ 7 (a) $9\mathbf{i} + 12\mathbf{j}$ (b) $4\frac{1}{2}\mathbf{i} + 6\mathbf{j}$ (c) $3\mathbf{i} + 4\mathbf{j}$
8 $-3\mathbf{i} + 8\mathbf{j}$, $-6\mathbf{i} + 16\mathbf{j}$, trapezium 9 (a) $8\mathbf{i} - 6\mathbf{j}$ (b) $0.6\mathbf{i} - 0.8\mathbf{j}$ (c) $-25\mathbf{i} + 60\mathbf{j}$
10 $18\mathbf{j}$ 11 $15\mathbf{i} - 20\mathbf{j}$ 12 $-4\mathbf{i}$ 13 $5\mathbf{i} - 2\mathbf{j}$ 14 $(3, -7)$, $6\,\mathrm{m\,s^{-1}}$,
15 $7\,\mathrm{m\,s^{-2}}$ 16 $9\sqrt{5}$, 153.4°

Exercise 4a (p. 31)

1 (a) 28.81, 29.29° (b) 8, 60° (c) 15, 53.13° (d) 27.17, 116.2°
2 62.72°   **3** 134.4°   **4** (a) 6.102 N, 131.0° (b) 6.842 N, 205.5°
(c) 14.14 N, 25° (d) 18.58 N, 327.7°   **5** 100, 86.60   **6** 2.113, 4.532   **7** 17.78 N, 137.0°
8 32.49, 11.43   **9** 35.07°, 5.202   **10** 292.8°, 31.87
11 17.87 N,   28.34°   **12** 126.9°   **13** 30°   **14** 22.25 N, 13.08 N   **15** 38.68°,
2.415 N   **16** 9.434 N

Exercise 4b (p. 37)

1 (a) 20 N, 34.64 N (b) 23.09 N, 19 N   **2** 36.87° (b) 50.19°
3 2.695 kg, 10.64 N   **4** 10.32 N, 12.21 N   **5** 60 N, 25 N
6 160 N, 120 N   **7** (a) 3.214 N, 3.830 N   (b) 4.195 N, 6.527 N
8 2.144 N, 5.856 N   **9** 2.5 N   **10** 6.946 N, 17.59 N
11 0.3571 N, 1.786 N   **12** 42.57 N, 35.44 N   **13** 24.46 N, 27.65 N
14 42.43 N, 30 N   **15** 48.59°, 2.646 N   **16** 26.91 N, 56.44°, 995 g
17 72.54°, 3.145 N, 4.253 N, 10.48 N.

Exercise 5a (p. 48)

1 (a) 80 J (b) 60 J   **2** (a) 560 J (b) 400 J   **3** 1028 J
4 (a) 50 J (b) 100 J   **5** 140 J   **6** 600 kJ   **7** 18.75 N
8 (a) 600 j (b) 360 J   **9** 120 J   **10** $1.89 \, \text{m s}^{-1}$ **11** $5 \, \text{m s}^{-1}$
12 $4.472 \, \text{m s}^{-1}$ **13** 41.41°   **14** 60°   **15** 0.3183 N
16 1.25 kg   **17** $6.633 \, \text{m s}^{-1}$ **18** 4 N   **19** 1050 kN
20 8.4 m   **21** 6.75 J   **22** 750 J   **23** $15 \, \text{m s}^{-1}$, $53\frac{1}{3} \, \text{cm}^2$.

Exercise 5b (p. 54)

1 1 kN   **2** 5 kw   **3** (a) $50 \, \text{m s}^{-1}$ (b) $31.25 \, \text{m s}^{-2}$ **4** 30 kW
5 $21.33 \, \text{m s}^{-1}$ **6** $30 \, \text{m s}^{-1}$ **7** 50 kN, 90 kN
8 2.8 kN, 3.8 kN, 38 kW   **9** $\frac{2}{15} \text{m s}^{-2}$ **10** 31.5 kW, $1.6 \, \text{m s}^{-2}$
11 $20 \, \text{m s}^{-1}$, 20 kN   **12** $0.25 \, \text{m s}^{-2}$, 500 N   **13** $58.48 \, \text{m s}^{-1}$
14 $4v$ **15** $6v$ **16** 1500 kW, $22.41 \, \text{m s}^{-1}$ **17** (a) $10Rv/3$ (b) $3W/7N$
18 $0.1 \, \text{m s}^{-2}$, 51 kW.

Exercise 6a (p. 61)

1 50 N cm, 110 N cm, 30 N cm   **2** 120 N cm   **3** 7 anticlockwise
4 165 N cm, 45 N cm, 30 N cm   **5** 19 clockwise   **6** 3, 4   **7** 5, 7
8 (a) 10 N, 12 cm (b) 2 N, 10 cm (c) 8 N, 4 cm (d) 3 N, 10 cm (e) $6F$, $8\frac{1}{2}a$
9 27 anticlockwise   **10** Clockwise couple of moment 17   **12** $4i$ at $(0, 3)$
13 $-6j$ at $(-3, 0)$ **14** $-3j$ at $(4, 0)$ **15** $16i$ at $(0, 1\frac{1}{2})$
16 7   **17** 9   **18** $-3, 4$

Exercise 6b (p. 66)

1 30 N, 40 N   **2** (a) 20 N, 40 N   (b) 120 N   **3** 75 cm   **4** 18 cm
5 (a) 120 N cm, 160 N cm (b) 20 N cm   **6** $4W$, $6W$, $5W$, $5W$
7 40 cm, 65 cm   **8** $5\frac{1}{2}a$, $1\frac{1}{2}a$ **9** 24 cm
10 (a) 30 N, 20 N   (b) 4 cm   (c) 200 N cm   **11** 6 kg, 70 cm   **12** 9, 15, 36
13 $3W$, $2W$, $3a$, Wa **14** $5W/3$, $4W/3$, $7a/5$, $W/7$, $15W/7$, W

Exercise 7 (p. 79)

1 $P\sqrt{34}$, 0.6, $4a/3$ from D on CD produced   **2** $\sqrt{61}$ N, 6.8 cm from A on CA produced
3 $2\sqrt{7}$ N, $\sqrt{3}/2$, 5 m from A on AC produced   **5** 22 N m

6 $\sqrt{149}$ N, 0.7, 8.286 cm from A on AB produced
7 $3y + 6x + 1 = 0$, $-3i + 6j$, 1 anticlockwise   **8** $20\sqrt{3}$ N m
9 $4\vec{AC}$ dividing AB in ratio $1:3$ **10** 10 N, $\frac{1}{3}$ m from C on BC produced
11 $1.2a$ from A, $4\sqrt{3}Fa$ **12** $2P\sqrt{21}$, $0.5669a$ **14** $P = S = -4$, $Q = 10$
15 $\sqrt{41}$, -0.8, $(-1,0)$ **17** $6\sqrt{3}$ N, $4\sqrt{3}$ N m   **18** 27.53 N m
19 (a) 60, 50 (b) 96, 50   **21** $4a$ from A on BA produced   **22** $y + 2x = 8$
24 $4i + 11j$, $(-\frac{1}{11}, 0)$ **25** 15 clockwise, $10i + 5j$ **26** $3y + 5x = 6$
27 (a) 10, $6\sqrt{3}$, (b) 2, $2\sqrt{3}$ **28** 10 N, 20 N m   **29** 40 anticlockwise, $-15i - 20j$
30 $5a/2$ from D on CD produced.

Exercise 8a (p. 86)

1 8 N s, 14 N s   **2** 30 N s, 53.24 N s   **3** 30 N s, 10 m s^{-1}
4 2 m s^{-1}, 30 J   **5** 28 m s^{-1}, 80 N s   **6** 33 m   **7** $Mmv^2/2(M+m)$
8 5.346 m s^{-1}, 10.69 Ns   **9** 8 N s, 400 m s^{-1},   320 kN   **10** 35.56 cm
11 540 kN   **12** 1.545 cm s^{-1} **13** 709.5 m s^{-1}
14 80 m s^{-1}, 96 kJ   **15** 200 m s^{-1}, 400 m s^{-1}.

Exercise 8b (p. 91)

1 $\frac{3}{4}$, 126 J   **2** $\frac{2}{3}$ **3** $\frac{3}{5}$ **4** $6\frac{1}{4}$ m   **5** 0.5 m s^{-1}, 2 m s^{-1}, 1.5 m s^{-1}
6 7 m s^{-1}, 9.5 m s^{-1}, 9 N s   **7** 0.8 m s^{-1}, 1.2 m s^{-1}, 19.2 J
8 17 m s^{-1}, 3 m s^{-1}, 32 N s   **9** $2/3$ **11** $2.2u$ **12** $7u$ due west, $6u$ due west
13 $\frac{1}{5}$ **14** $6u$, $8u$, $\frac{1}{3}$ **15** $7u/9$ **16** 3, $13u/64$, $15u/64$, $9u/16$
17 2, $u/9$, $20u/81$, $32u/81$ **18** 5 s   **19** 5 s   **20** $(3e-1)/(e+1)$ **21** $e < 0.38$
22 33/49   **24** $u/4$, $3u/4$, $1/2$ **25** 12 s

Exercise 9a (p. 97)

1 27.36 N, 75.18 N, $20°$ **2** 6.5 N, 2.5 N
3 $41.81°$, 24.49 N at $72.28°$ to the horizontal   **4** 9.326 N, 22.07 N
5 21.36 N, 7.5 N   **6** $0.9165W$, $23.58°$ **7** 3 N, 4 N
8 (a) 13.66 N, 12.25 N (b) 6.830 N, 8.851 N
10 0.2887 N, 0.2887 N at $60°$ to the horizontal
11 2.617 N, 8.417 N at $71.88°$ to the horizontal
12 5.422 N at $67.24°$ to the horizontal   **13** $a/2$, $a/4$ **14** 49 cm
15 0.6385 N, 0.8208 N, 72 cm   **16** $4a\sqrt{3}/3$.

Exercise 9b (p. 108)

1 0.4663   **2** 12.99 N, 51.29 N   **3** 2.887   **4** 61.54 N
5 0.3530   **6** 200 N   **7** 20.98 N   **8** 1   **9** 0.5445
10 2   **11** 0.5959   **12** $3r$ **13** $30.96°$ **14** 21.86 N
15 0.7892 of length of ladder   **16** $2W$, $(3W \cot \theta)/4$, $(3 \cot \theta)/8$
17 $\frac{1}{2}W(1 + \sin^2 \theta)$, $\frac{1}{2}W \sin \theta \cos \theta$, $(\sin \theta \cos \theta)/(1 + \sin^2 \theta)$ **19** 7/24
20 11.04 N, 18.12 cm   **21** $l < 2\mu h$ **22** $14Wa$ **23** 0.5774
24 Topples. Force $= 0.6820W$ **25** 0.2247
26 A: $13W/8$, $21W/32$; B: $11W/8$, $21W/32$. $\mu_{\min} = \frac{21}{44}$ **27** $\frac{24}{43}$
28 1.323 W at $10.89°$ to the horizontal.

Exercise 10 (p. 116)

1 2 m s^{-2}, 20 N, 10 N, 4 N   **2** 5 m s^{-2}, 150 N, 300 N   **3** 7.5 m s^{-2}, 52.5 N, 30 N
4 2.5 m s^{-2}, 37.5 N, 75 N   **5** 4 m s^{-2}, 48 N, $48\sqrt{2}$ N   **6** 6 m s^{-2}, 48 N, $48\sqrt{2}$ N
7 2 m s^{-2}, 12 N   **8** 5 m s^{-2}, 1.5 N   **9** $\frac{1}{9}$ m
10 4 m s^{-2}, 72 N, $72\sqrt{3}$ N   **11** 2 m s^{-2}, 80 N, 104 N   **12** 0.7 s
13 $4\frac{2}{3}$, $1\frac{1}{3}$ **14** 3 m s^{-2}, 130 N, 350 N   **15** $3\frac{1}{3}$ m s^{-2}, 4 N
16 1.707 s   **17** 0.4 s, 0.2 s   **18** $4mg/3$, $3m$ **19** 2.45 m s^{-1}
20 $3.6mg$ **21** 3.062 m s^{-2}, 123.9 N   **22** Upwards, $\frac{2}{3}$ m s^{-2}, 160 N, $266\frac{2}{3}$ N
23 1.5 m below.

Exercise 11a (p. 125)

1	1.118 m s^{-1}	**2**	50.66 cm	**3**	76 cm	**4**	11.43 cm
5	33.56°	**6**	25.33 cm	**7**	1.033 m s^{-1}	**8**	2.739 m s^{-1}
9	88.42 cm	**10**	6.333 cm	**11**	2:9	**12**	$\sqrt{3ag}$, $2mg$
14	3.619 cm	**15**	4.099 m s^{-1}	**16**	20 cm	**17**	6.957 cm
18	15.81 rad s^{-1}	**19**	3.464 m s^{-1}, 2449 m s^{-1}.				

Exercise 11b (p. 130)

1	800 kN, arc tan $\frac{1}{3}$	**2**	116 cm	**3**	12.5 cm	**4**	1.5 m, $\frac{1}{2}$
5	8/9	**6**	28.14 m s^{-1}	**7**	37.42 m s^{-1}	**8**	135.6 m s^{-1}
9	13.75 m s^{-1}	**10**	arc tan $\frac{5}{12}$, 34.31 m s^{-1}.				

Exercise 12 (p. 140)

1 8.324 km h^{-1} on a bearing of 19.86°   **2** 30 m s^{-1} on a bearing of 81.87°
3 18.03 km h^{-1} on a bearing of 213.7°   **4** 5i − j   **5** − 2i + 3j
6 24.57 m s^{-1} on a bearing of 276.8°   **7** 130 m s^{-1} on a bearing of 337.4°
8 (a) 6.325 m s^{-1} (b) 15 m   **9** 51.77°   **10** 5 s
11 (a) 10.92 m s^{-1} on a bearing of 76.19° (b) 119.4 m
12 (a) 10j, 4 m (b) − i + j, 2.121 m   **13** 4.222 h   **14** 39.16 cm s^{-1}
15 1.414 km, 5.657 km   **16** 12 m, 1 s   **17** 48 m   **18** 1.964 km, 5.518 min
19 (a) 6 (b) 7 (c) 5   **20** A bearing of 41.81°, 53.67 m, 80.50 m
21 12.07 m s^{-1} on a bearing of 118.0°   **22** 56.60 m, 152.0°   **23** 12.59 s
24 4 h 15 min   **25** 4.950 m s^{-1} on a bearing of 224.4°
26 Due east, 20 km h^{-1}, 34.64 km h^{-1} **27** 2.5 m s^{-1} **28** Bearings of 300° ± 41.41°
29 98.13°   **30** 65 m, 53.13° ± 22.62°

Exercise 13a (p. 146)

1	5i − 12j	**2**	3i + 2.75j	**3**	12 m	**4**	9.372 m
5	4 s, 69.28 m	**6**	23.32 m s^{-1}, 59.04°	**7**	20i + 15j		
8	15 m s^{-1}	**9**	(a) 0.5 s, 0.7 s	(b) 2.078 m		**10**	20 m s^{-1}, 53.13°
11	40.36°	**12**	16i + 9j	**13**	4 m	**14**	18.34 m
15	0.8 s, 1.6 s	**16**	5i + 9j	**17**	1.875 s, 9.75 m	**18**	11.36
19	6.009 m s^{-1}	**20**	108 m, 57.6 m	**21**	1.5 m	**22**	5 $\sqrt{2}$ m s^{-1}.

Exercise 13b (p. 153)

1	34.64 m, 40 m	**2**	57.49°	**3**	2.121 s	**4**	60 m
5	25.69°, 64.31°	**6**	22.36 m s^{-1}	**7**	10 m	**8**	51.96 m
9	arc tan 4	**10**	240 m	**11**	39.69 m		
12	30 m s^{-1}, 22.5 m	**13**	tan 15°:1	**14**	arc tan $\frac{1}{2}$, arc tan $1\frac{1}{2}$		
16	18 m **17** 3.248 m	**19**	1 m **20** $\frac{3}{4}$ **21** $\frac{8}{9}$.				

Exercise 14a (p. 157)

1	1 m, 1.35 m	**2**	0.05, − 1.45	**3**	17a/30	**4**	0.8 cm, 6.4 cm
5	7 m, 4 m, $\sqrt{65}$ m	**6**	2.8 m	**8**	13	**9**	5 m
10	9.5, 3.5						

Exercise 14b (p. 163)

1	4.789 cm	**2**	4.2 cm	**3**	8$\frac{8}{9}$ cm, 7$\frac{7}{9}$ cm, arc tan $\frac{7}{8}$		
4	7 cm, 10 cm	**5**	arc tan $\frac{5}{7}$	**6**	9$\frac{10}{11}$ cm, 63.23° **7** 1$\frac{1}{3}$ kg		
8	12.37 cm	**9**	arc tan $\frac{4}{21}$, 0.2219 N	**10**	0.3h **11** arc tan $\frac{2}{3}$		
12	arc tan $\frac{9}{8}$, 0.045 N	**13**	13a/15	**15**	2a, Mg $\sqrt{2}/6$ **16** 6.425 cm		
17	90° + arc sin $\frac{1}{3}$	**18**	arc tan $\frac{10}{9}$.				

Exercise 14c (p. 171)

1 $(\frac{4}{5}, \frac{2}{7})$ 2 $(\frac{45}{28}, \frac{93}{70})$ 3 $(1, -\frac{2}{3})$ 4 $(\frac{3}{4}, \frac{12}{5})$ 5 $(0.3, 1.5)$
6 $(\frac{2}{5}, 0)$ 7 $\bar{x} = 1/(e-1),\ \bar{y} = (e+1)/4$ 10 arc tan $3/2\pi$
11 $d\sqrt{6}/6$ 12 $\frac{2}{3}$ 13 $\frac{45}{28}$ 14 $1\frac{2}{3}$ 15 $\frac{4}{5}$
16 $(e^2 + 1)/2(e^2 - 1)$ 17 arc tan $\frac{1}{16}$ 18 arc tan $\frac{16}{11}$
19 $27a/40,\ 7a/40$ 20 $R\sqrt{6}/2$ 21 arc tan $\frac{29}{17}$
22 arc tan $\frac{1}{4}$ 25 arc tan $\frac{5}{9}$

Exercise 15a (p. 178)

1 (a) $\frac{1}{26}$ (b) $\frac{11}{13}$ (c) $\frac{4}{13}$ (d) $\frac{1}{2}$ 2 (a) $\frac{3}{8}$ (b) $\frac{1}{2}$ 3 (a) $\frac{1}{12}$ (b) $\frac{5}{36}$ (c) $\frac{1}{6}$ (d) $\frac{11}{12}$
4 (a) $\frac{1}{16}$ (b) $\frac{1}{4}$ 5 (a) $\frac{1}{6}$ (b) $\frac{1}{12}$ (c) $\frac{1}{2}$ 6 (a) $\frac{3}{8}$ (b) $\frac{3}{4}$
7 (a) $\frac{1}{5}$ (b) $\frac{1}{15}$ (c) $\frac{4}{5}$ 8 (a) $\frac{1}{27}$ (b) $\frac{19}{27}$ 9 $\frac{65}{81}$ 10 (a) $\frac{1}{6}$ (b) $\frac{29}{30}$
11 (a) $\frac{9}{55}$ (b) $\frac{16}{495}$ 12 (a) $(1-x)^3$ (b) $1-x^3$ 13 (a) $\frac{1}{5}$ (b) $\frac{3}{25}$
14 (a) $\frac{1}{6}$ (b) $\frac{1}{3}$ (c) $\frac{2}{3}$ (d) $\frac{1}{20}$ (e) $\frac{3}{5}$ 15 (a) $\frac{1}{2}$ (b) $\frac{1}{6}$ 16 (a) $\frac{1}{4}$ (b) $\frac{7}{20}$ (c) $\frac{21}{190}$
17 (a) 0.73 (b) 0.66 (c) 0.59 (d) 7 18 (a) 200 (b) 2800 19 (a) $\frac{6}{7}$ (b) $\frac{30}{49}$
20 (a) $\frac{1}{120}$ (b) $\frac{1}{12}$ (c) $\frac{1}{10}$ 21 (a) $\frac{13}{24}$ (b) $\frac{3}{4}$ (c) $\frac{5}{92}$ 22 (a) 3 (b) 9

Exercise 15b (p. 184)

1 (a) $\frac{7}{12}$ (b) $\frac{3}{4}$ 2 $\frac{21}{25}$ 3 (a) $\frac{3}{8}$ (b) $\frac{7}{13}$ 4 (a) $\frac{17}{20}$ (b) $\frac{3}{5}$ 5 $\frac{3}{50}$, Yes
6 (a) $\frac{5}{6}$ (b) $\frac{3}{16}$ 7 $\frac{13}{24}$ 8 $\frac{8}{15}$ 9 Yes 10 (a) No (b) Yes (c) No
11 $\frac{5}{24}$ 12 $\frac{5}{18}$ 13 $\frac{11}{20}$ 14 (a) $\frac{3}{8}$ (b) $\frac{3}{8}$ (c) $\frac{5}{16}$ 15 $\frac{13}{15}$
16 (a) $\frac{3}{8}$ (b) $\frac{27}{64}$ 17 (a) No (b) Yes (c) Yes (d) No 18 $\frac{3}{20}$, No 19 $\frac{16}{27}$
20 0.6 21 (a) Yes (b) No (c) Yes 22 0.11 23 Yes
24 (a) $\frac{1}{8}$, Yes (b) $\frac{1}{14}$, No (c) $\frac{3}{28}$, No

Exercise 15c (p. 189)

1 (a) $\frac{7}{15}$ (b) $\frac{14}{45}$ (c) $\frac{2}{9}$ 2 (a) 0.666 (b) 0.538 3 (a) $\frac{15}{28}$ (b) $\frac{3}{7}$ 4 (a) $\frac{4}{9}$ (b) $\frac{7}{18}$
5 (a) $\frac{39}{125}$ (b) $\frac{4}{25}$ (c) $\frac{96}{125}$ 6 (a) $\frac{7}{18}$ (b) $\frac{1}{3}$ (c) $\frac{17}{36}$ 7 (a) $\frac{5}{12}$ (b) $\frac{1}{3}$
8 (a) $\frac{2}{3}$ (b) $\frac{1}{2}$ 9 (a) $\frac{2}{9}$ (b) $\frac{4}{7}$ 10 $\frac{9}{13}$ 11 (a) $\frac{3}{4}$ (b) $\frac{5}{13}$
12 (a) $\frac{1}{21}$ (b) $\frac{1}{2}$ 13 $\frac{5}{9}$ 14 $\frac{76}{91}$

Exercise 16a (p. 195)

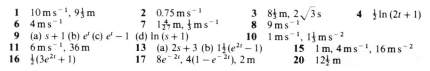

1 $10\,\text{m s}^{-1}, 9\frac{1}{3}\,\text{m}$ 2 $0.75\,\text{m s}^{-1}$ 3 $8\frac{1}{3}\,\text{m}, 2\sqrt{3}\,\text{s}$ 4 $\frac{1}{2}\ln(2t+1)$
6 $4\,\text{m s}^{-1}$ 7 $1\frac{4}{27}\,\text{m}, \frac{1}{3}\,\text{m s}^{-1}$ 8 $9\,\text{m s}^{-1}$
9 (a) $s+1$ (b) e^t (c) $e^t - 1$ (d) $\ln(s+1)$ 10 $1\,\text{m s}^{-1}, 1\frac{1}{3}\,\text{m s}^{-2}$
11 $6\,\text{m s}^{-1}, 36\,\text{m}$ 13 (a) $2s+3$ (b) $1\frac{1}{2}(e^{2t}-1)$ 15 $1\,\text{m}, 4\,\text{m s}^{-1}, 16\,\text{m s}^{-2}$
16 $\frac{1}{2}(3e^{2t}+1)$ 17 $8e^{-2t}, 4(1-e^{-2t}), 2\,\text{m}$ 20 $12\frac{1}{2}\,\text{m}$

Exercise 16b (p. 200)

1 (a) $3t^2\mathbf{i} - 2\mathbf{j}, 6t\mathbf{i}$ (b) $12\mathbf{i} - 2\mathbf{j}, 12\mathbf{i}$
2 (a) $(t^2 + 2)\mathbf{i} + (t^3/3 + 3)\mathbf{j}$ (b) (11, 12) (c) $(6, 5\frac{2}{3})$ 3 $8y + x^2 = 0$
4 $3\sqrt{5}, -\arctan\frac{1}{2}$ 5 $8y = 3x^2, 2\sqrt{10}$ 6 $(6, 2\frac{2}{3}), 2\sqrt{5}$
7 (a) (1, 2), 1 s (b) $3\mathbf{i} + \mathbf{j}$ 8 (a) $(-1, -1), 2\sqrt{10}$ (b) $\frac{1}{2}, 2\sqrt{3}/3$
9 2 11 5 12 (a) $2\sqrt{5}$, (b) $18\sqrt{5}$ 13 $\sqrt{17}$ 14 8

Exercise 17a (p. 207)

1 $1.571\,\text{m s}^{-1}, 1\,\text{m s}^{-1}, 4.935\,\text{m s}^{-2}$ 2 (a) $1\frac{1}{3}\,\text{m s}^{-1}$, (b) $1.920\,\text{m s}^{-1}$, (c) 1 m
3 (a) $0.8019\,\text{m s}^{-2}$ (b) $1.021\,\text{m s}^{-1}$ 4 $1.306\,\text{m}, 3.245\,\text{s}$
5 $3\,\text{s}$ 6 $2\,\text{m}, 4.189\,\text{s}$ 7 $1.5\,\text{m}, 3.512\,\text{m s}^{-1}$
8 $1.111\,\text{m s}^{-1}$ 9 $2.216\,\text{m}, 8.285\,\text{m s}^{-1}$ 10 $2.424\,\text{m s}^{-2}$

Exercise 17a (*Contd.*)

11 $1.5\,\mathrm{m\,s^{-1}}$	**12** $3.819\,\mathrm{m\,s^{-1}}$	**13** $3.820\,\mathrm{m}$	**14** $2.084\,\mathrm{m}$
15 $0.7071\,\mathrm{m}$	**16** $4.636\,\mathrm{m\,s^{-1}}$	**17** 54%	
18 $a\sqrt{3}/2,\ a\pi/T$	**19** $0.6a$	**20** $T/6$	**21** $V/2\pi f.$

Exercise 17b (p. 215)

1 (a) $3.441\,\mathrm{s}$ (b) $91.29\,\mathrm{cm\,s^{-1}}$ **2** $9.733\,\mathrm{m\,s^{-2}}$
3 $25\,\mathrm{cm}$ (a) $1.897\,\mathrm{m\,s^{-1}}$ (b) $1.049\,\mathrm{m\,s^{-1}}$ **5** (a) $1.405\,\mathrm{s}$ (b) $1\,\mathrm{m\,s^{-1}}$ (c) $0.1881\,\mathrm{s}$
6 $6.333\,\mathrm{cm}$ **7** $1.662\,\mathrm{s},\ 0.1587\,\mathrm{m}$ **8** $0.4443\,\mathrm{s}$
9 (a) $0.6283\,\mathrm{s}$ (b) $7.5\,\mathrm{m\,s^{-2}}$ **10** $2.766\,\mathrm{s}$ **11** $24.85\,\mathrm{cm\,s^{-1}}$
12 $90.66\,\mathrm{cm\,s^{-1}}$ **14** $\tfrac{1}{2}\sqrt{3eg}$ **15** $4\sqrt{eg/5},\ 4g/5$
16 $64.62\,\mathrm{cm\,s^{-1}}$ **17** (a) $2.598\,\mathrm{m\,s^{-1}}$ (b) $35\,\mathrm{N}$ (c) $0.3846\,\mathrm{s}$
19 (a) $0.2484\,\mathrm{s}$ (b) $0.3501\,\mathrm{s}$ **20** $0.6050\,\mathrm{s}$ **21** $0.7493\,\mathrm{s}$
22 $1.257\,\mathrm{m\,s^{-1}},\ 29\,\mathrm{cm}.$

Answers to Revision Exercises

Exercise R1 (p. 217)

1 $4\,\mathrm{m\,s^{-2}}, 22\,\mathrm{m\,s^{-1}}$ **2** $2\,\mathrm{s}, 31.25\,\mathrm{m}$ **3** $4\,\mathrm{s}$ **4** $16\,\mathrm{m}, 1\,\mathrm{s}$ **5** $2\frac{1}{2}\,\mathrm{m\,s^{-2}}$
6 $2\,\mathrm{m\,s^{-2}}, 3\,\mathrm{m\,s^{-1}}$ **7** $2.4\,\mathrm{s}$ **8** $2\frac{1}{4}\,\mathrm{m\,s^{-2}}$ **9** $2\,\mathrm{m\,s^{-2}}$ **10** $8\,\mathrm{s}$

Exercise R2 (p. 218)

1 $8\,\mathrm{m\,s^{-2}}$ (b) $6\,\mathrm{kg}$ **2** $280\,\mathrm{N}$ **3** $16\,\mathrm{N}$ **4** $2W$ **5** $20\,\mathrm{N}$ **6** $3x/g$
7 (a) $1.25W$ (b) $0.2W$ **8** (a) $5\,\mathrm{m\,s^{-2}}$ (b) $37\frac{1}{2}\,\mathrm{N}$ **9** $125\,\mathrm{N}$
10 $48\,\mathrm{N}, 30\,\mathrm{N}$ **11** $32\,\mathrm{mg}, 52\,\mathrm{mg}$ **12** (a) $3.2\,\mathrm{N}$ (b) $2.8\,\mathrm{N}$
13 (a) $5\,\mathrm{m\,s^{-2}}$ (b) $60\,\mathrm{N}$ (c) $10\,\mathrm{N}$ **14** (a) $2g$ (b) $45W$ (c) $9W$ **15** $15\,\mathrm{N}$

Exercise R3 (p. 219)

1 (a) $\overrightarrow{DA}$ (b) $\overrightarrow{AC}$ (c) $\overrightarrow{BA}$ (d) $\overrightarrow{BC}$ **2** (a) $6\mathbf{i} - 2\mathbf{j}$ (b) $-8\mathbf{i} + 4\mathbf{j}$ (c) $-\mathbf{i} + 5\mathbf{j}$
(d) $-6\mathbf{i} + 8\mathbf{j}$ (e) $2\mathbf{i} - 6\mathbf{j}$ **3** (a) $25, 73.74°$ (b) $\sqrt{5}, 116.6°$ (c) $20, -53.13°$
(d) $4\sqrt{10}, -161.6°$ **4** $-3\mathbf{i} - 10\mathbf{j}$, parallelogram   **5** (a) $96\mathbf{i} + 28\mathbf{j}$ (b) $15\mathbf{i} - 36\mathbf{j}$
(c) $-0.6\mathbf{i} + 0.8\mathbf{j}$ (d) $\mathbf{i} + \sqrt{3}\mathbf{j}$ **6** (a) $\pm(3\mathbf{i} - 4\mathbf{j})$ (b) $\pm(75\mathbf{i} + 40\mathbf{j})$
7 $24\mathbf{i} - 18\mathbf{j}$ **8** $16\mathbf{i} - 8\mathbf{j}$ **9** $36\mathbf{i} - 21\mathbf{j}$ **10** $-10\mathbf{i} + 3\mathbf{j}$
11 $(27, -36), 18\mathbf{i} - 24\mathbf{j}$ **12** $5\mathbf{i} - 9\mathbf{j}$ **13** $12\mathbf{i} + 17\mathbf{j}$ **14** $43, 87$

Exercise R4 (p. 219)

1 (a) $11.40\,\mathrm{N}$ (b) $7.549\,\mathrm{N}$ (c) $26\,\mathrm{N}$ (d) $17.62\,\mathrm{N}$ **2** $73.40°$ **3** $136.6°$
6 (a) $6.033\,\mathrm{N}, 40.65°$ (b) $6.883\,\mathrm{N}, 65°$ (c) $17.22\,\mathrm{N}, 235.9°$ **7** $16, 13.86$
8 $50.35\,\mathrm{N}, 78.32\,\mathrm{N}$ **9** (a) $15\,\mathrm{N}, 7\,\mathrm{N}$ (b) $32\,\mathrm{N}, 11.2\,\mathrm{N}$ **10** $21.67, 14.99$
11 $132.8°$ **12** $14\,\mathrm{N}, 48\,\mathrm{N}$ **13** (a) $7.937\,\mathrm{N}, 41.41°$ (b) $10.39\,\mathrm{N}, 60°$
14 $61.28, 53.89$ **15** $9.460\,\mathrm{N}, 16.51\,\mathrm{N}$ **16** $44.26\,\mathrm{N}, 8.816\,\mathrm{N}$ **17** $74.39\,\mathrm{N}, 48.29\,\mathrm{N}$
18 $23.94\,\mathrm{N}, 18.62\,\mathrm{N}, 16.38\,\mathrm{N}$ **19** $16.32\,\mathrm{N}, 12.94\,\mathrm{N}, 25\,\mathrm{N}, 7.139\,\mathrm{N}$

Exercise R5 (p. 221)

1 (a) $1080\,\mathrm{J}$ (b) $720\,\mathrm{J}$ **2** (a) $2\,\mathrm{m\,s^{-1}}$ (b) $43.53°$ **3** (a) $75\,\mathrm{kW}$ (b) $0.5\,\mathrm{m\,s^{-2}}$
4 $65\,\mathrm{kN}, 0.575\,\mathrm{m\,s^{-2}}$ **5** $700\,\mathrm{W}$ **6** (a) $6.928\,\mathrm{m\,s^{-1}}$ (b) $26.46\,\mathrm{cm}$ **7** $1.5\,\mathrm{m\,s^{-2}}$
8 (a) $40\,\mathrm{m\,s^{-1}}$ (b) $20.80\,\mathrm{m\,s^{-1}}$ **9** $104\,\mathrm{N}$ **10** $360\,\mathrm{N}$ **11** $26\,\mathrm{kW}$
12 (a) $1.292\,\mathrm{m\,s^{-2}}$ (b) $31.75\,\mathrm{m\,s^{-1}}$ **13** (a) $2v$ (b) $v/2$ **14** $20\,\mathrm{cm}$ **15** $1.5\,\mathrm{kg}$

Exercise R6 (p. 222)

1 $24\,\mathrm{N\,cm}, 78\,\mathrm{N\,cm}$ **2** $6\,\mathrm{N}, 4\frac{2}{3}\,\mathrm{cm}$ **3** (a) $60\,\mathrm{N}, 80\,\mathrm{N}$ (b) $52\frac{1}{2}\,\mathrm{N}$ (c) $700\,\mathrm{N\,cm}$
4 $-2, -6$ **5** (a) $22\,\mathrm{cm}$ (b) $54\,\mathrm{cm}$ **6** $6, 3$ **7** $8, -2$ **8** (a) $5\frac{1}{2}a$ (b) $7W, 3W$
9 (a) $5\mathbf{i}$ at $(0, -7)$ (b) $8\mathbf{j}$ at $(-9, 0)$ **10** $5\mathbf{i}$ at $(0, -2)$ **11** $60\,\mathrm{N}$
12 $30\,\mathrm{cm}, 20\,\mathrm{N}, 50\,\mathrm{N}$ **13** $4, -9$ **14** $14\,\mathrm{N}, 12\,\mathrm{N}, 8\,\mathrm{N}, 18\,\mathrm{N}, 80\,\mathrm{N\,cm}$

Exercise R7 (p. 223)

1 $\sqrt{41}$ N, $1\frac{1}{4}$, 2 cm from A **2**   13 N, 2.4, $4\frac{1}{2}$ cm from A on BA produced
3   (a) 20.69 cm from A, (b) 14.64 cm from A **4**   (16,0), $(0, -4)$ **5**   6.607 N, 6.359 cm
6 $4\sqrt{5}$, $y = 2x + 3$ **7**   18 N cm   **8**   6.385 N   **10** -12, 6, 25   **11**   8, 4, $6\sqrt{3}$
12 $2\sqrt{2}$, $45°$, (0,8)   **13** $2\sqrt{13}$, 8 cm from A, 10 N cm   **14** $8\mathbf{i} - 6\mathbf{j}$, $4y + 3x = 26$
15   10, 3   **16**   (a) 8, -1 (b) -12, 15
17   (a) 15 N, -10 N, 42 N cm   (b) -15 N, 30 N, -138 N cm

Exercise R8 (p. 225)

1   6 m s^{-1}, 240 J   **2**   800 m s^{-1}, 120 N s   **3**   80 cm   **4**   120 cm   **5**   3 s
6   6 m s^{-1}, 8 m s^{-1}, 12 J, 6 N s   **7**   2 m s^{-1}, 5 m s^{-1}, 45 J, 10 N s   **8**   37.5 g
9   8 m s^{-1}, 1 m s^{-1} **10**   20 m s^{-1} **11**   3 m s^{-1}, 13 m s^{-1}, $\frac{5}{7}$ **12**   20 m s^{-1}
13 $\frac{74}{27}$, $\frac{84}{27}$, $\frac{4u}{9}$ **14**   12 cm s^{-1}, $\frac{3}{4}$ **16**   5 s   **17** $\frac{3}{32}$ **18** $e < \frac{1}{2}$

Exercise R9 (p. 226)

2 $\frac{3}{4}$, 35 N   **3**   1.363   **4**   44.14 N, 18.65 N   **5**   13.74 kg   **7**   225.2 N, 0.2887
8   300 g, 43.29 N, 42.34 N   **9**   60.13 N, 35.83 N   **10**   0.5594
11   (a) 22.36 N, 22.36 N along AC; (b) 28.28 N, 20 N along DA **12**   0.2750
13   1, 3   **14**   1.608 N, 6.212 N, 19.65 cm   **15** $\dfrac{W}{2}$
16 $2W\cos\theta$, $\dfrac{3W - 2W\cos^2\theta}{\sin\theta}$, $\dfrac{2 - \cos 2\theta}{\sin 2\theta}$ **17** $\dfrac{W\sqrt{3}}{2}$, horizontal, $\dfrac{\sqrt{3}}{2}$ **18** $\frac{1}{3}$, $\frac{1}{4}$

Exercise R10 (p. 228)

1   6 m s^{-2}, 160 N, 320 N   **2**   5 m s^{-2}, 45 N   **3**   4 m s^{-2}, 3.6 N, 5.091 N
4   3.5 m s^{-2}, 1 N   **5**   72 N, 60 N, 48 N   **6**   250 N
7   (a) 125 (b) 750   **8**   450 cm   **9**   2.2 s   **10**   3 s   **11**   10 kg
12   (a) 9 kg (b) 25 kg (c) 5 kg   **13** $1\frac{1}{2}$ s   **14**   (a) $\frac{1}{4}$ (b) 220 kg

Exercise R11 (p. 229)

1   (a) 4.189 rad s^{-1} (b) 2.5 rad s^{-1} **2**   6 m s^{-1}, 8 rad s^{-1} **3**   1.125 m
4   4.082 rad s^{-1} **5**   (a) 274.2 N (b) 4.056 m s^{-1} **7**   115.2 N, 58.61°
8   1.792 m s^{-1} **9**   9.129 rad s^{-1} **10**   48 cm   **11**   2.291 m s^{-1}
12   5.797 m s^{-1}, 150 N   **13**   (a) 9.258 rad s^{-1} (b) 6.325 rad s^{-1} **14**   23.66 m s^{-1}, $\frac{14}{15}$
15   25.46 m s^{-1} **16**   50 m s^{-1}, 15.08 m s^{-1} **17**   0.3734

Exercise R12 (p. 230)

1   102 km h^{-1} on a bearing of 28.07°   **2**   (a) $8\mathbf{i} - 6\mathbf{j}$ (b) 10 m s^{-1}
3   5.142 km h^{-1} on a bearing of 230°   **4**   25 m s^{-1}, 100 m   **5** $-10\mathbf{i} + 15\mathbf{j}$
6   (a) 200 s (b) 120 s   **7**   (a) 2 (b) 6 (c) 2.828
8 $1\frac{1}{2}$ **9** $(-2,0)$ **10** (a) 10 s (b) 17.04 s   **11**   3.465 h
12   (a) 1.512 min (b) 150 m   **13**   60 m, 208.8 m   **14**   (a) 240 m (b) 298.2 m
15   (a) 3.830 km h^{-1} (b) 4.981 km h^{-1} **16**   10°, 8.595 km h^{-1}, 48.75 km h^{-1}
17   (a) $5\frac{1}{2}$ s (b) 1.789 s   **18**   123.7 m   **19**   (a) 1.5 m s^{-1} (b) 3.324 m s^{-1}
20 $(11, -2)$, $(5, -10)$

Exercise R13 (p. 232)

1 $12\mathbf{i} + 5\mathbf{j}$, 13 m s^{-1}, 22.62°   **2**   3 s, 60 m   **3**   6.984 s, 59.42 m, 199.4 m
4   (a) 62.5 m (b) 3.536 s   **5**   24.30°, 65.70°   **6** $28\mathbf{i} + 16\mathbf{j}$, $56\mathbf{i} + 12\mathbf{j}$
7   (a) 53.13° (b) 35 m s^{-1} **8** $1\frac{1}{2}$ s, $3\frac{1}{2}$ s   **9**   33.75 m   **10**   (a) 6 s   (b) 69.28 m

11 (a) 54.74° (b) 15° or 75° (c) 30° **12** arc tan $\frac{1}{2}$, arc tan $2\frac{1}{2}$
14 30.47 m s^{-1}, 60.26° **15** 70.4 m, 22 m **16** 12 m s^{-1}, 20 m **17** 2 m, 6 m
18 $\frac{1}{2}$ **19** arc tan 3 **21** 24**i** + 11**j** **22** $u\sqrt{13}$, u^2/g

Exercise R14 (p. 234)

1 (−1, 3) **2** 16, −18 **3** 12$\frac{1}{4}$ cm **4** (a) 7 cm, 6$\frac{2}{3}$ cm (b) 58.78° **5** 5.713 cm
6 (a) 1.789 N (b) 3.354 N **7** (a) 3$\frac{1}{3}$N (b) 2 N **10** 2$\frac{8}{9}$g **11** 15.64°
12 (a) (1.5, 0.9) (b) (1, $\pi/8$) **13** arc tan $\frac{14}{17}$ **14** 2:3

Exercise R15 (p. 235)

1 (a) $\frac{1}{24}$ (b) $\frac{1}{52}$ (c) $\frac{1}{2}$ **2** (a) $\frac{1}{10}$ (b) $\frac{3}{20}$ (c) $\frac{9}{20}$ **3** $\frac{5}{6}$, $\frac{17}{18}$ **4** 98/125
5 (a) $\frac{13}{30}$ (b) $\frac{12}{145}$ **6** (a) $\frac{1}{12}$ (b) $\frac{1}{6}$ (c) $\frac{1}{3}$ **7** $\frac{11}{24}$ **8** (a) $\frac{1}{10}$ (b) $\frac{1}{15}$
9 (a) $\frac{23}{54}$ (b) $\frac{4}{9}$ (c) $\frac{5}{18}$ **10** (a) $\frac{13}{25}$ (b) $\frac{144}{625}$ (c) $\frac{48}{625}$ **11** $\frac{3}{5}$, $\frac{1}{6}$
12 (a) $\frac{1}{35}$ (b) $\frac{18}{35}$ **13** (a) $\frac{21}{128}$ (b) $\frac{35}{128}$ **14** (a) Yes (b) No (c) Yes (d) No (e) Yes
15 (a) $\frac{33}{50}$ (b) $\frac{17}{100}$ **16** (a) $\frac{1}{24}$, No **17** (a) $\frac{3}{32}$ (b) $\frac{3}{128}$ **18** (a) 0.36 (b) 0.47
19 (a) $\frac{5}{11}$ (b) $\frac{2}{13}$ (c) $\frac{4}{33}$ **20** (a) $\frac{15}{23}$ (b) $\frac{2}{3}$
21 $\frac{5}{18}$ **22** $\frac{2}{3}$ **23** $\frac{3}{5}$ **24** $\frac{4}{13}$

Exercise R16 (p. 237)

1 (a) 4 m s^{-1} (b) 3 s (c) 6$\frac{1}{4}$ m s^{-1} **2** (a) (4,0) (b) 16 s
3 (a) 5 m s^{-1} *at* arc tan $\frac{4}{3}$ to the positive *x*-axis (b) $9y = (x + 6)^2$ **4** $\frac{7}{8}$ s, 4 m
5 (a) 5 N (b) 1 N, 2$\frac{1}{2}$ N **6** (a) 7.616 m s^{-1} (b) (2,0)
7 6 m s^{-1} **9** (a) 6.403 m s^{-1} (b) 2.712 m
10 (a) 8**i** + 5**j**, 8**i** + 10**j** (b) 13 m s^{-1} (c) $\frac{1}{2}$ s **11** (a) 4 s (b) 6 m s^{-1}
12 (a) 4 s (b) 3 s **13** (a) (8, −5$\frac{1}{3}$) (b) (26, 5$\frac{1}{3}$) **15** (36, 18)
16 $y^2 = x + 3$, $y + 2x = 15$, (6,3) **17** $\frac{1}{8}$ s **18** 12 s, 213 m
21 (a) $\pi/6$ s (b) $5\pi/6$ s (c) 3 m, 1 m

Exercise R17 (p. 239)

1 1.257 m s^{-1}, 1.974 m s^{-2}, 1.165 m s^{-1} **2** 1.212 m, 1.466 m s^{-1}
3 (a) 1.5 m, πs (b) 1.910 m s^{-1} (c) 2.524 m s^{-1} **4** 1.402 m, 2.057 s
5 2.472 m s^{-1} **6** (a) 71.39 cm (b) 1.058 m s^{-1} **7** $T/6$, $5T/12$ **9** (a) 2.177 s
(b) 2.199 m s^{-1} **10** (a) 55.41 cm s^{-1} (b) 52.24 cm s^{-1} **11** 94 **12** (a) 11.26 cm
(b) 3.197 cm **13** (a) 9.952 cm (b) 0.4527 s **15** 0.5447 s

the millinery w/hs
84-87 south gate rd
NJ JJ F